SLEEP GROOVE

Andrews McMeel Publishing
a division of Andrews McMeel Universal
1130 Walnut Street, Kansas City, Missouri 64106

www.andrewsmcmeel.com

25 26 27 28 29 LAK 10 9 8 7 6 5 4 3 2 1
ISBN: 978-1-5248-9295-1
Library of Congress Control Number: 2024946909

Editor: Lucas Wetzel
Art Director: Diane Marsh
Production Editor: Jennifer Straub
Production Manager: Julie Skalla
Assistant Production Editor: Kelsey Rolofson

ATTENTION: SCHOOLS AND BUSINESSES
Andrews McMeel books are available at
quantity discounts with bulk purchase for
educational, business, or sales promotional
use. For information, please e-mail the
Andrews McMeel Publishing Special Sales
Department: sales@andrewsmcmeel.com

SLEEP GROOVE

WHY YOUR BODY'S CLOCK IS SO MESSED UP AND WHAT TO DO ABOUT IT

OLIVIA WALCH, PHD

Andrews McMeel
PUBLISHING®

CONTENTS

FOREWORD

WE OFTEN THINK ABOUT sleep as a simple game of addition and subtraction.

Six hours of sleep per night, five days a week, plus ten hours of sleep per night on weekends? That's more than seven hours of sleep per night.

Midnight to 8:00 AM one night and 9:00 PM to 5:00 AM the next? Those are two eight-hour blocks of equally good sleep.

But the truth is that sleep is a dynamic process, not a counting exercise. And the dynamics of sleep and circadian rhythms—as we shift our sleep timing night-to-night, jet-travel across time zones, work overnight shifts, or watch Netflix into the wee hours—are a lot more complicated than adding and subtracting.

Just because they're more complicated, though, doesn't mean you can't come to understand them.

I'll never forget the fateful day I first met Olivia Walch, PhD. The setting: a seventh-floor conference room in the University of Michigan's main hospital. The players: me, an enthusiastic young neurologist specializing in sleep medicine; her, a graduate student in applied mathematics. The topic at hand: how ideas from physics could help us to track our body's internal clocks, a.k.a. our circadian rhythms.

As a graduate of the Northwestern Sleep Medicine fellowship with a math degree of my own, I felt like someone had just told me the sky was purple. Circadian rhythms are famously hard to measure, which is one of the reasons they often get short shrift when we talk about sleep. But that day, Olivia explained how mathematical approaches could be used to track circadian rhythms in the real world. This, in turn, could help people sleep better in the real world by targeting and boosting their rhythms. It was an area primed for combining her background in mathematics and wearables with my background in sleep medicine—and the start of a beautiful friendship.

In the years that followed, Olivia and I would work together on a wide range of projects, from elite athletic performance (Go Blue!) to chronic fatigue in multiple sclerosis patients, all aimed at leveraging *rhythms* to improve real-world sleep. Along the way, she built up an arsenal of analogies for how rhythms matter for sleep, capturing the ideas in the models she used without needing to bring in a single equation.

More importantly, Olivia became a part of our family, bringing my (non-mathematical) husband up to speed on AI over dinner and teaching my son and his third-grade classmates how the internal body clock works with her amazing cartoons. Her love for explaining (and cartoons) comes through in this book.

As you embark on the journey in the pages that follow, set aside your preconceptions about sleep duration as "the" definition of healthy sleep, and get curious about the growing body of research showing that healthy rhythms are foundational to our sleep and overall well-being. Start thinking about the timing of your sleep and not just a running tally of hours you've slept. There's a lot that can be gained from living in synchrony with the clocks that time us—and you don't need any fancy math to feel the difference for yourself.

Cathy Goldstein, MD
Professor of Neurology at the University of Michigan Sleep Disorders Center

INTRODUCTION

YOU PROBABLY KNOW ABOUT eight hours. I knew about eight hours, back during my worst days of not sleeping (college). The thinking I indulged in at the time went a little something like:

> ME: I'll sleep from 1:30 AM to 4:30 AM tonight, 2:30 AM to 6:30 AM tomorrow
>
> ME: and 8:00 PM tomorrow to 1:00 PM the day after, which is eight hours a night
>
> ME: *on average*

There were many reasons why this was a bad idea. For starters, the fact that I was short on sleep nearly all the time made me terrible at learning. I'd be taking notes during a lecture when my handwriting would start cramping in on itself, getting smaller and more cursed-looking with every word. *You're drifting off,* I'd think, beaming the thought from my brain like an escape pod from a collapsing planet. *Stab yourself in the leg with your pen.*

Which never worked. Neither did pinching myself, holding my feet off the ground, or alternating sips of black coffee and blue raspberry slushie (Hot Drink/ Cold Drink™). Eat a single Cheerio at a time and I'd reliably stay awake during class— only to pass out on the communal laundry room floor fourteen hours later, halfway through both the problem set due the next day and the load of clothes I needed to wear when turning it in.

That's fine, I'd think. *I'll catch up on the weekend.*

Cut to Saturday, when I'd wake up at 1:00 PM after seventeen hours of sleep and still feel tired.

Lest you think I was losing all this sleep in college because I was cool, let me state clearly: No, I was just a try-hard. I was double-majoring in try-hard disciplines [1] and taking probably too many credit hours a semester while also trying to do every club. I *did* have fun doing it, and I did learn along the way, but, wow, did I waste time (my professors', my own). I would have gotten so much more out of those classes—and would remember now so much more of the fun I had back then—if I'd just been thinking about sleep in a better way.

The elements were all there, in retrospect. I was taking neuroscience classes and thinking about the brain at the same time as I was taking physics and thinking about the mechanics of how things move. You don't need calculus to have a pretty good idea of how a single pendulum is going to swing back and forth, or how a projectile is going to fly through the air, or how a ball is going to bounce. It's intuitive. Live long enough on Earth, and you just sort of get it.

So when I'd finish a physics problem, I'd do a sanity check for my work: *Does it make sense that the basketball would bounce two thousand miles into the air? No? Okay, let's check to see if you made any mistakes around the time you were stabbing yourself in the forearm.*

But I never thought about the brain, and sleep specifically, as something that could have a straightforward, intuitive physics of its own. The brain seemed too complicated for that: a dark, unknowable blob that we could only understand in pieces, chunks lopped off one at a time and given long names for med students to memorize.

The irony is that in implicitly deciding that the brain was too complicated to understand like a physics problem, I ended up thinking about sleep in an incredibly simplistic way. Sleep was like grades or my bank account or bean counting: a tally where any shortfall could be rescued in the nick of time by a deposit.

It turns out that, yes, while the brain is complicated, so too is that basketball: It's composed of countless molecules, it's experiencing air resistance and friction, and there are all sorts of other forces and considerations and factors in play that can make describing it fully a massive headache. But that doesn't mean you can't step back and squint and have a pretty decent idea of how it's going to bounce.

And when you step back and squint at sleep, you see it's a rhythm.

---

1   *mathematics, biophysics*

I don't mean anything fancy by that. Cycling between sleep and wake from day to day is a rhythm the same way breathing is a rhythm, or your heartbeat, or walking, or swinging on a swing at a playground. It's a thing that repeats itself, where one loop through the whole cycle tends to take about as long as the ones before and after.

But unlike breathing and the beating of your heart, where the rhythm part is front and center to our understanding of what it means to be healthy, sleep's rhythm is typically an afterthought, pushed aside in favor of its duration. "Healthy" sleep, we're told, is hitting a target number of hours of sleep per night (say, eight), without much focus on when those hours happen. And when we don't hit that target number, we can just make up for it—on the weekend, on the couch watching TV, whenever.

The focus on how *much* you're sleeping over the more general *how* you're sleeping (and how it fits within the rest of your day) reduces sleep health to a single number for us to track, hack, and obsess over. We worry when a wearable says it's lower than it should be, even when we feel perfectly fine, or we brag about needing less of it to get by than everyone else.

Imagine treating other rhythms in your body the way we treat sleep: *Oh, it's no problem if my heart stops for the next thirty seconds—I'll make up for it by having it beat twice as fast right after.* [2] Or: *Maybe you need to breathe fifteen times a minute, but I can get by just fine with ten.* [3]

If I played you two heartbeats—one with a clear, robust rhythm and the other scraggly and erratic—you'd never say they were equally healthy just because they had the same number of beats in a minute.

Yet, when it comes to our sleep, we tend to think that two sleep/wake patterns are equally healthy so long as they have a good amount of sleep (eight-ish hours) a night, on average.

BACK TO ME

So I knew about eight hours in college and did decently well at hitting that target with weekend sleep binges, but I was tired all the time, barely forming any new memories and sleeping on the laundry room floor. What changed?

A couple of things. First, I went to graduate school and learned about the math of sleep and circadian rhythms, a.k.a. your body's internal clock. I'm not going to focus too much on the technical details in this book, but I am going to pull out more analogies from

---

2   *Cool. Good for you, bro.*

3   *Cool. Cool.*

physics and elsewhere to help build an intuitive sense—like we have for how gravity works—of how your body wants to move in and out of sleep over the course of the day.

Next, I was a subject in a sleep research study myself, in which I had to keep a consistent bedtime and wake time every single night for three months. For ninety days, I stuck with an 11:30 PM bedtime and a 7:30 AM wake time. I wore a sleep tracker that would rat me out to the researchers if I broke the rules. No exceptions. No late nights. If I didn't get my work done, too bad—11:30 PM was lights out.

It's hard to overstate how much participating in this study affected my life. There is basically no aspect of my health and quality of life that was left unchanged during that period: My skin cleared up, my mood improved, I was the fittest I'd ever been in my life, and I earned a cool thirty-five dollars as compensation for being a subject. I finished that study and never looked back. My sleep, to this day, is still basically the same as it was back then: really, really regular.

Of course, this is an N = 1 result. I have no concrete proof that, for me, it was the regular schedule that led to all the good things. Plus, people are different, and life constraints are different, and keeping a consistent schedule of any kind can be close to impossible if you've got young kids, or a job where you work shifts, or a sleep disorder, or construction outside.

But the reason I've made the choice to dedicate a huge chunk of my life to spreading the word about sleep regularity and circadian rhythms—through this book, through my role as a researcher in the sleep field, and through the sleep/ circadian tech start-up I founded—is because of how utterly transformative I believe getting it right was for me. There's a clear *before* and after in my life, where the dividing line is "that study I signed up for that made me change my sleep for three months." And it's a lot better here on the *after* side of things.

WHY "GROOVE"?

If I were to describe what was different for me after that study, I might say that by the time it wrapped up, I'd found a sleep groove. A groove, here, is an effortless rhythm with momentum. It feels good, and you don't need to think about it. Heartbeats aren't the best comparison point here because we don't tend to consciously think about them, but walking, jogging, dancing—all of these have rhythms that can just start to click.

You're not in a groove if you're carefully tiptoeing around broken glass on the sidewalk. You are in a groove if you're striding down the street while winking and making finger guns at your neighbors, like an overconfident protagonist about to fall from grace at the start of a movie.

You're not in a groove if you're sitting on a swing at a playground, half-heartedly stubbing your shoe into the mulch as you lurch forward and backward a few inches. You are in a groove if you've pumped your legs back and forth enough to get some serious height, and now it all feels easy, and if you jumped off at the crest of the swing you could actually hurt yourself, that's how high and fast you're going.

Signs of a sleep groove include: Falling into sleep quickly and easily at the time you want to. Waking up without an alarm. Feeling alert during the day. Feeling resilient to nights where your sleep isn't so good. Feeling like there's momentum to how you fall into and emerge from sleep.

This is how I've slept for the last decade, and if it doesn't line up with how you're sleeping now, I hope this book can help you get to something like it.

WHAT TO EXPECT AHEAD

Speaking of the last decade: over the course of my years thinking about this stuff, I've arrived at a few general principles for talking about sleep. They'll be sprinkled throughout the other chapters of this book, but I want to lay them out here to set the stage for what's to come. The first one, which is probably the biggest, is:

YOU CAN'T TALK ABOUT SLEEP WITHOUT TALKING ABOUT CIRCADIAN RHYTHMS

Talking about sleep without talking about circadian rhythms is like trying to shoot an arrow at a target without paying attention to the wind direction. Sure, sometimes it works out. But other times, the fact that you're ignoring a critically important (yet mostly invisible) factor means you end up way, way off course. You end up sleep-bingeing, like I did back in the day, and wake up still feeling bad.

There's a problem here, which is that most people only kind of know what circadian rhythms are. That was me for a long time. I was probably two years into studying it as a research topic before the definition of a "circadian rhythm" really clicked for me. The definition I learned for a circadian rhythm was:

> *Any pattern in the body which repeats approximately every twenty-four hours, is endogenously generated, can entrain to a new schedule, and is temperature-compensated.*

What that means is "anything in your body that repeats itself about once a day, would continue to repeat itself in the absence of all external stimuli, could adapt itself to new patterns of stimuli, and still keeps decent track of time even when your body heats up and cools down." **4**

Which is a lot of things! Your grip strength, your immune response, your core body temperature, your metabolism, your production of the hormones melatonin and cortisol—all of these are rhythms that are "circadian."

With so many things falling under that definition, it's hard to briefly and effectively express what circadian rhythms *are*, which results in people tending to think that circadian rhythms and sleep are the same thing (they're not), or that circadian rhythms are like sleep's little cousin (no).

Circadian rhythms are the missing piece to so many of sleep's grand mysteries in day-to-day life. Getting in bed and feeling tired but not able to fall asleep? Waking up at 3:00 AM and not knowing why. Playing roulette with a nap that could be either forty-five minutes or six hours? So often, these phenomena are circadian rhythms making themselves known.

Wondering about them without considering the potential role of circadian rhythms is like practicing archery during a cyclone and wondering why your aim is off.

And on the topic of rhythms:

SLEEP RHYTHMICITY NEEDS TO BE PRIORITIZED
THE SAME WAY WE PRIORITIZE SLEEP DURATION

Sleep duration—how long you sleep—definitely matters: I would have been even more of a mess on four hours of sleep every night in college than I was with my irregular eight-hour average. But it's by no means the whole story.

More and more research has shown that the *timing and regularity* of sleep can be as important as sleep duration, if not more so, especially if you're getting more than six hours of sleep a night. This means interrogating the notion that you can and should always use the weekends to catch up on some sleep. It also means putting aside the idea of sleep as something fungible. The second you make the choice to stay up late, you've thrown off the rhythm of your sleep, and paying it back isn't as simple as just sleeping longer the next day. If anything, paying it back in that way could disrupt your rhythms even more.

But before "even one bad night is enough to disrupt your sleep rhythms" strikes fear into your heart, let me also say:

---

4 *This last one is particularly impressive since the "gears" of our body's clock are molecules bumping into each other, and molecules bump into each other faster as temperature rises. Marvel at this biological wonder which is 1) extremely cool and 2) not going to appear anywhere else in this book.*

WE FOCUS TOO MUCH ON SLEEP FACTORS OUTSIDE OUR CONTROL AND NOT ENOUGH ON SLEEP-RELATED FACTORS WITHIN OUR CONTROL

Nobody can just get in bed and command themselves to fall asleep. At least, almost nobody. Every time I say this in public, there's always one person who abruptly brakes their car from two lanes over, rolls down the window, and yells, "ACTUALLY I CAN FALL ASLEEP ANYWHERE, ANYTIME." [5]

In general, though, the ability to fall asleep at will is not one possessed by the majority of the population. And achieving a low-stress, blissful descent into sleep is only made harder by the fact that a lot of the media coverage on sleep will tell you that every minute you spend failing to direct yourself into unconsciousness is shortening your lifespan, increasing your risk of cardiovascular disease, scarring your brain, and blunting your relationships with your loved ones.

Which isn't to say that the research behind that media isn't good or right. I'm grateful for all the work that's gone into understanding the short-term pains of lost sleep, the long-term health consequences of a lifetime spent sleeping poorly, and the mechanisms and meaning of different sleep stages. But none of those insights are particularly helpful to a person who's trying to sleep in the moment, nor do they always touch on mechanisms that are under our direct control.

Nobody can go back in time and change the way they slept when they were twenty-one. Nobody slides under the covers, tracksuit on, and says, "I'm gonna REM sleep so hard tonight." You can't make yourself fall asleep on command, and you especially can't make yourself fall into a sleep stage of your choosing through willpower alone.

So when articles talk about how your past sleep habits could be affecting you now, or when devices, apps, and wearables give so much attention to the fraction of each night you spend in each sleep stage, they emphasize things outside of your control to the detriment of things *within* your control. [6]

Things like light exposure. Exercise. Activity during the day. Caffeine intake. Meal timing. Trying cognitive behavioral therapy for insomnia.

---

5   *Good for them.*

6   *To be clear, I'm still going to talk a lot about how sleep duration and other sleep numbers relate to health outcomes. But I'm also going to avoid engaging with questions like "Am I going to die if I don't get seven hours of sleep tonight?" in favor of "Am I doing everything I can to set myself up for sleep success? Yes? Great. That's all I can do."*

Out of all of these, I feel like light exposure is the one that's the most disproportionately undervalued for as much as it matters to our sleep. Which is a shame, because:

WE LIVE IN A LIGHT-POISONED WORLD

Two hundred years ago, if you wanted to stay up late, you could do it. You'd just be in the dark the whole time. And sure, not *dark*-dark—I'll allow you a candle and some torches. But the environment would be dim.

This dimness matters because our body's circadian rhythms are set by alternating patterns of light and dark, activity and inactivity. You need *both* parts of the pattern to get into a groove.

Think of your circadian rhythms like a person on a swing . . .

. . . and light exposure as a shove in the forward direction:

When you're swinging forward, a push in that direction is going to speed you up or make it so you swing higher. You get a faster swing or more of a swing, a bigger *amplitude*.

But if you're on the backswing, a shove in the forward direction is the exact thing you don't want. It's going to slow you down and crush your swing's momentum. It's like this:

Light during the day is that push in the good direction that speeds you up and boosts your groove. Light at night is that push in the wrong direction—that smacking-into-someone-when-you-don't-expect-them-to-be-in-the-way sensation.

And in modern life, we get a *lot* of light at night. From high-efficiency LED streetlights to bright fluorescent overhead lights in dorms and apartments, we're splattered by the stuff from dusk to dawn. We're dosing ourselves with light around the clock, and it's squashing our circadian rhythms the same way that a jerk standing behind us while we're trying to enjoy ourselves at a playground would squash the height of our swing.

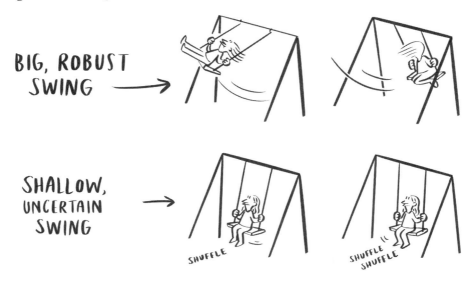

BIG, ROBUST SWING →

SHALLOW, UNCERTAIN SWING →

SHUFFLE

SHUFFLE SHUFFLE

Ever since humans got control of the light switch, we've lost the clear differential we used to have between day and night. And it's the loss of that differential that is the real problem, because:

LIGHT ITSELF ISN'T BAD

One reason I like that swing analogy is because it makes this point quite nicely. You *want* light exposure when it's giving you a big shove in the right direction. Light at that time is actively good! Light exposure during the day builds our brain's confidence that it's daytime. It boosts our mood, makes it easier to fall asleep at bedtime, and makes us more robust to the effects of light at night. [7]

---

7   *Exceptions include: you've traveled recently, you've been on a weird schedule lately, you're a shift worker, or you've got some type of circadian condition like non-24 going on. This will come up over and over again throughout the book, so I might not bring it up every time. But concepts like "morning"/"day" or "evening"/"night" will always have an implicit asterisk of *not always, because the time we care about is* <u>biological time,</u> *not* <u>wall-clock time,</u> *and biological time and wall-clock time can be very different things.*

Avoiding light during the day—by staying inside or wearing those very clear-looking glasses that purport to block blue light—is very likely the opposite of what you want to do if your aim is to improve your sleep. You should be soaking it up as much as you can since getting light during the day is one half of the "clear-cut difference between day and night" equation. It's only later that light switches from sleep-promoting to sleep-toxic. The timing is the key.

And it's not just timing light. Everything I said about light here can also be said about activity, meals, and other signals that send timing cues to your body. Round-the-clock living has made it so we get a constant barrage of confusing, contradictory signals at the wrong times that throw off our collective circadian groove.

The good news is that for most people it's not all that complicated to figure out how to time things better. Building up your intuition for the *physics of sleep* will help you get a feel for how your light exposure and other behaviors are boosting or messing with your groove. That's what this book is for.

A COUPLE DISCLAIMERS

I'm not going to talk about sleep stages (they're cool, just not my scene). I'm mostly going to skip over sleep disorders, like apnea and narcolepsy (I'm a mathematician, not a health provider).

I'm going to use an unholy number of analogies and metaphors. You'll think, *No. Surely. Surely not another metaphor,* and then I'll pull out "sleep is kinda like an A.C. unit," and you'll drop your head into your hands in despair. I love analogies for communicating this kind of stuff. It gives you the gist of what our current scientific understanding is, and if a fellow scientist wants to (probably correctly) argue that what I said isn't exactly right, I can pull out a long pointing stick with chalk at the end and circle the word "kinda."

I'm going to minimize paragraphs of the type: "Researchers discovered (sleep thing) correlates with (bad thing), (bad thing), (bad thing), and even (bad thing) . . ." Nothing against paragraphs of that kind, but so often the interesting stuff of a scientific paper is in the *details*, not the soundbite you can tweeze out of the title. Plus, to be a bit repetitive here, the more robust a sense you can build for *why* something might be good or bad, the less you need some specific study finding some specific correlation with a good or bad thing to motivate a change in your behaviors. I'd rather you have an instinctive feel for what a sleep groove means vs. having a Rolodex of facts about how bad sleep is killing you.

Some of the chapters here have been adapted from blog posts I've written for my company blog, which is another disclaimer: I run a tech company called Arcascope and have a financial incentive (that goes beyond "selling more books") to get you to think about sleep the way I do. My hope is that this book leaves you half as interested in the awesome stuff we're building at Arcascope as I am. [8]

WHAT TO DO IF YOU'RE THE KIND OF PERSON WHO LOSES INTEREST HALFWAY THROUGH BOOKS

Go to bed at the same time every night, as much as you can. Give yourself at least two weeks of strict adherence to see how this changes things for you. Keep in mind that it can take a while for your sleep rhythm to find the beat.

Get bright daylight during the day and the darkest dark you can get at night. Be super aware of your light exposure over the course of the day, especially overhead lights in the evening. Don't nap during the day if you're having trouble falling asleep at night, but also don't do dangerous things like driving while you're super tired. Don't do work in bed. If you're lying in bed but unable to sleep, get up and move somewhere else so you don't start to associate "bed" with "not sleeping." Keep the lights down as you do it, but not so dark that you trip and hurt yourself.

And if none of this is helping you, reach out to a sleep doctor about a diagnostic sleep study or cognitive behavioral therapy for insomnia.

FOR EVERYONE ELSE

(*stage whisper*) Keep going . . . !!

---

8   *10% as interested is also fine*

WHAT IF I JUST
STOP SLEEPING?

LET'S START WITH THE most obvious example of a non-grooving sleep groove there is: not sleeping. Who says you even need to? Sure, *I, personally*, was miserable on my nights with low sleep, but aren't there people out there who can get by with four hours a night? How do you *know, scientifically*, that one hour of sleep a day would make a person miserable?

The answer is that we (sleep scientists) have put people on schedules like that before, and those people are generally pretty miserable. They get sick and feel bad and become really cranky. Not saying you couldn't be the exception. If you are, good for you.

YOU'RE WELCOME TO TRY, BUDDY

Unrelatedly, in 1933, a young man walked into the New York State Psychiatric Institute, told the researchers there he didn't think humans needed to sleep, and offered them the chance to study him as he gave it up entirely in one grand push. The gentleman, dubbed "Z" by the scientists, was twenty-four years old and "somewhat eccentric" in behavior and dress. His expectation, he told them, was that it would be a bit tough at the beginning, but that he'd power through and arrive on the other side newly freed from sleep's treacherous clutch on his daily hours.

"This is a terrible idea," said the scientists. "But, if you insist, practice typing on this." Then they shoved a typewriter into his hands, which could be used to monitor his transcribing skills over time as a proxy for alertness. [9]

He insisted. They weren't equipped to keep him in a lab for the duration of the study, but he agreed to come in regularly for check-ins so they could monitor his progress. To make sure he wasn't falling asleep when nobody was looking, they gave him a "a watchman's recording clock and key" and instructed him to turn the clock's key every ten minutes to prove he was awake at that time. [10] Then he went home and set off on his mission to break his pesky sleep habit once and for all.

Z actually did . . . pretty well, all things considered? His first accidental sleep occurred only at the two-day point, and he pulled off sixty-two hours of continuous consciousness on day six. Sure, he was still sleeping accidentally, sometimes for more than an hour at a time, but in terms of reclaiming hours of his day for wakefulness, his overall performance was not shabby at all.

Yet the quality of those reclaimed hours became increasingly terrible. The investigators had planned to give him said typing error test throughout the duration of the study, but by day four, he couldn't manage it—focusing on the page he was attempting to transcribe was too hard and his eyes hurt too badly. If he went into a dark room or closed his eyes a hair too long, he immediately fell asleep.

He was irritable and hallucinating, mistaking a desk for a water fountain and losing track of which building he was in. He was also paranoid and unreasonable, convinced that one of the experimenters was specifically out to get him [11] and erupting into rants when he didn't get his way.

Determined as he was to prove human sleep was optional, he repeatedly insisted this was all normal behavior for him.

---

9 *What they actually wrote: "For several reasons we tried to discourage him, but as he insisted, we directed him to practice writing on a typewriter for a period of thirty minutes daily, marking the typing at the end of each minute of work. We also suggested that he practice a code substitution test. These tasks, if thoroughly mastered, should offer a measure of the effect of sleeplessness on the mental functional ability."*

10 *In a sense, this is one of the earliest sleep trackers, a sort of steampunk Fitbit.*

11 *"He became more and more certain that this experimenter was personally interested in making life disagreeable for him and in interpreting his behavior in terms of pathologic mental mechanisms. These tendencies became so marked during the last two days of the vigil that they formed the principal basis for our decision to discontinue the experiment when we did."*

"Sure," said the researchers, presumably scribbling on their clipboards. "Whatever you say." [12]

By day ten, his grudge against his enemy on the experimental team was so bad and disruptive that the researchers decided to wrap the whole thing up.

When you read the paper about this experiment, the conclusion is much less focused on the fact that he became a combative paranoid jerk and much more focused on the fact that his brain didn't melt. Which is true! He showed no drop-off in "intelligence tests" given to him on the second and seventh days ("It is probable that the tests were too easy," admit the scientists), his motor function seemed fine throughout, and the vital signals they measured were all in normal ranges.

The conclusion of their paper reads:

> In general, it has been demonstrated that in this case it was possible to go with practically no sleep for approximately ten days without any known physiologic effect and without any permanent change in the personality or in the mental function.

There's a key word in the above: "practically." After all, he did sleep.

A THING ON THE INTERNET IS WRONG ABOUT SOMETHING

There's a decently well-known creepypasta called "The Russian Sleep Experiment" online. By "decently well-known," I mean that it 1) has, at least right now, its own Wikipedia page translated into thirteen languages but 2) is a creepypasta, a thing that many readers might never have heard of before. For those folks: a creepypasta is a spooky story on the internet. Now we're all on the same page.

The plot of "The Russian Sleep Experiment" is basically that a bunch of people are kept artificially awake for fifteen days thanks to a mysterious-yet-horrifying experimental gas. When the experimenters go in to check on them at the end, they learn (scary spoiler alert) that they've been *devouring their own organs since day five of the trial.*

When I read this, I thought, "Self-cannibalism after only five days? Weak."

Let me now explain why.

---

12 *"He stated that these instances, which we regarded as evidence of disorientation, were more or less usual occurrences in his everyday behavior. This may be true, but we believe that the disorientation was exaggerated beyond that which he habitually showed."*

WHY I DON'T THINK YOU'D START EATING YOURSELF AFTER FIVE DAYS WITHOUT SLEEP

For starters, purposefully subjecting yourself to extreme sleep deprivation is really not so rare a feat. People were doing it all the time in the middle of the last century. A radio DJ named Peter Tripp famously stayed up and on the air for 201 hours as part of a stunt in 1959. Intriguingly, there were reports that his personality changed after the experience, which, while much milder than what happens in the fictional "Russian Sleep Experiment," certainly sparks the imagination. Does extreme sleep deprivation damage something permanently inside us? *Does it shatter our minds for good?*

Well, maybe, but the evidence isn't particularly overwhelming. If anything, efforts to track down the answers to this question have mostly led to answers like, "Eh, probably not?"

Take, for instance, a 1968 study looking to answer the question of "Does acute sleep deprivation induce permanent psychosis?" Unlike with Z's attempt to stop sleeping completely in 1933, where he was at home for large swaths of time and accidentally fell asleep at multiple points, this study intended to rigorously ensure no sleep was had by anyone at any point by keeping the participants under observation the entire time. No caffeine or other drugs would be allowed, but bright lights, loud sounds, and other environmental stimuli would be permitted.

Four participants signed up for the study: a college student, another college student, a hippie, and a young man who it later turned out was on the run from the FBI. As with Z, they started falling asleep pretty quickly in their sleep loss journey, but this time they had round-the-clock study coordinators on hand to shake them awake when it happened. When the study coordinators got tired of shaking everyone awake all the time (they, too, were losing sleep to work night shifts on the study), the researchers told the participants that the rules for compensation had changed: if any one of them fell asleep, nobody was getting paid the four hundred dollars they had been promised.

So the participants started shaking each other awake and dunking their heads in ice water to keep conscious. They also got snippier with each other and more paranoid, gossiping to the researchers about who they thought was going to make it. Around day five, they experienced something called the "fifth day turning point"—observed in other extreme sleep loss studies—where the participants simultaneously felt better about the whole not-sleeping thing and also started having wild, vivid hallucinations.

One of them hallucinated seeing Humpty Dumpty and the image of a gorilla merging and coming toward him menacingly. He (understandably) panicked and begged to be taken out of the study. "We hear you when you say you want to leave the study," said the researchers. "But have you considered *not* wanting to leave the study?" [13]

"You're right," said the subject, after several minutes of being coaxed not to leave by the scientists and other participants. "I've whipped it." Then he rejoined the study and proceeded to stay awake with the other participants for a total of 205 hours, or 8.5 days, at which point the researchers decided they weren't learning anything new and shut the whole thing down.

The participants were allowed to sleep and kept around for observation for three recovery days before being released back out into the world. By the time they were leaving the lab, the participants did report feeling like they had been changed by the experience, largely for the better. Three of them had grown beards. One of the students said the experience had given him a new sense of focus. "I am never doing drugs again," said the hippie. "Nothing compares to the high of sleeplessness." [14]

Six months later, when they checked in with everyone to look for long-term effects, the hippie was a "confirmed 'acid head'" living in Haight-Ashbury. The students were still students. It was around this time that they learned that the remaining subject was on the run from the FBI for deserting the Navy and suspected homosexuality.

"Great," said the researchers. "Nobody is more psychotic than when they started." [15]

I have the detachment of time and distance, which makes elements of this study genuinely funny to me, [16] but I'm also horrified at the thought of research

---

13 *What they actually report happening is this: "He then sobbed bitterly and talked about his father . . . [His reaction] was explained to him as a reactivation of infantile fears and conflicts which appeared in the form of his old night terrors because of the fatigue and stress of seven days of sleeplessness . . . After a few minutes, he asked that he not be taken from the experiment 'because of a little thing like that.'"*

14 *What actually happened: "He was quite proud of himself for having mastered the ordeal and he made frequent comparisons between sleep deprivation and his experiences with LSD . . . [h] e claimed that sleep deprivation had given him insight which he had not attained with LSD and he doubted that he would ever use LSD again."*

15 *What they actually said: "Thus, it seems that the experiment was merely a brief interlude in the lives of the four subjects. It did not change the directions they were going nor modify their life styles."*

16 *The fact that the authors included the line "confirmed 'acid head'" in their paper is one of them.*

with an explicit goal of determining if sleep deprivation makes you permanently psychotic. *What if the answer had been yes? How could learning that possibly be worth it?* I'm also unable to imagine running a research study where we wouldn't let a subject go home as soon as they said they wanted to be let out (unless there was a risk that they weren't safe to drive, in which case we'd let them out, then strongly encourage them to nap first in a spare room before leaving). Protecting human subjects isn't something we can ever become complacent on, but it's at least good to know things have changed since this 1968 study was run.

BUT THIS ISN'T EVEN THE SLEEP DEPRIVATION RECORD!

Despite how unhappy the subjects were staying up for 205 hours, that's still more than two days short of a record set a few years earlier by a high schooler experimenting on himself for his local science fair over Christmas break.

Randy Gardner, age seventeen and described by the scientists who studied him as a healthy, cool, popular guy, [17] decided to stay awake for 264 hours because he'd been "fascinated by 'extremes' since early adolescence" and because 260 hours was in the *Guinness Book of World Records*.

First, two friends were the ones watching him to make sure he stayed awake, then the local press found out, then some scientists got wind of it and drove down from Stanford to tape electrodes to his face, and the whole thing ended up becoming national news. He was closely monitored throughout, took no stimulants to stay conscious, and sat for a psychiatric interview at 262 hours with the researchers.

The findings? Apart from one hallucination at ninety hours and a general sense of annoyance at people asking him nonstop him if he was awake, he was pretty much fine.

Quoth the researchers: "On the night before the final day, the subject wandered about the city in the company of one of the authors and one of his companions. There was little in his behavior during this period . . . to suggest that he had been awake longer than his companions." They also made sure to note in their write-up that he was still good at pinball, right up to the end.

He passed out for nearly fifteen hours the first night he returned to sleeping, but woke up on his own. After that, he went back to school. A week later, he was sleeping about seven hours a night.

---

17   *"[A] confident, self-assured individual," they write in one paper. "He makes friends easily, and usually has one or two close friends."*

"Yep," said the researchers. "Not more psychotic than when he started."[18]

It's truly remarkable how "pretty much fine" Randy was during the experiment. Curiously fine, the same way that Yao Ming[19] is curiously taller than me. Given that he *chose* to stay awake as his way of achieving an extreme, it's entirely possible that Randy Gardner is at the extreme end of the bell curve when it comes to the ability to withstand sleep deprivation and self-selected into an experiment he was likely to thrive in. There's nothing wrong with that, though it should probably make us a bit cautious about extrapolating his experience to others. Taking him as a benchmark for what the rest of us can do sleep-wise could be a bit like taking Yao Ming as a benchmark for what the rest of us should be able to reach off of high shelves.

Yet, in both the 1968 study and in Randy's world record–breaking science fair project,[20] people who presented as healthy seemed to come out the other side about the same as they went in.

Still, okay, fine—264 hours is eleven days. There are higher numbers of days a person could stay awake. The *Guinness Book of World Records* has stopped monitoring the record for Longest Time Awake because they're worried about people hurting themselves in the process. But if you went for it anyway . . .

DO YOU DIE?

My gut feeling is that if you went around to people and asked them, "If you weren't allowed to ever sleep, ever, would you eventually die?" the most common answer would be "Yeah, probably." That'd be my answer too, with an asterisk of "not that we've ever seen it happen for sure."

Said more strongly, there's no evidence for "a lack of sleep" being the thing to take a human out. There are reports of people operating with little to no sleep for weeks and then tragically dying, but in those cases it's impossible to prove that the sleep deprivation was the thing holding the knife in the shadows. Maybe they had an underlying heart condition? Maybe they had a vulnerable immune system and the sleep deprivation weakened it to the point where a serious infection could take hold? Maybe something else was going on and the not-sleeping was a symptom of that?

---

18 *"This suggests that the psychosis of sleep deprivation is not a general phenomenon inevitably brought on by prolonged wakefulness, and that other factors should be considered in explaining frequent development of psychosis observed by others."*

19 7'6"

20 *which did, in fact, win first place at the San Diego Science Fair that year*

It's *great* that we don't know for certain that this could happen because it means we've never run a controlled clinical trial where we take two people, completely deprive one of them of sleep while keeping everything else the same between the two of them, and then wait to see if the sleepless one dies.

We have done this with mice. One of the ways to sleep deprive a mouse (or other rodent) is "gentle handling." Gentle handling is exactly what it sounds like: the researcher watches the mouse, and if they notice it looks like it's about to sleep (or if an electroencephalogram [EEG] hooked up to its brain starts to show sleep-like patterns), they give the mouse a lil' shake. Maybe they nudge the cage; maybe they poke the mouse a little bit with their hand or brush. This requires having a trained researcher on hand and paying attention to the mice the whole time, and it's hard to standardize from person to person (how do you make sure everyone nudges with the same amount of gentleness?), but it's positively quaint compared to the other ways you can keep rodents awake.

These are the *automated* ways of ensuring a mouse doesn't sleep, the most established of which is called the disk-over-water (DOW) technique, introduced initially in 1983. I'm of two minds on disk-over-water: On the one hand, I believe fully in the necessity of experimental research with lab animals to forward our understanding of health for the greater benefit of all. On the other hand, it reads a bit like the kind of contraption a villain would try to use to take out 1960s James Bond.

Here's how it works:

- Take two rodents
- Put both of them in cages, divided but next to each other
- Make the floor for both cages a single disk
- Put a pool of (shallow) water underneath
- Hook up an EEG to both rodents to continuously monitor their sleep/ wake status
- Deem one of the rodents to be the "No Sleep" rodent. (The other is the "Control" rodent)
- If the No Sleep rodent's EEG waves start to look like it's falling asleep, automatically start spinning the floor, forcing both rodents to move if they don't want to be forced into the water

NOTE: The Control rodent can fall asleep whenever it wants, but it will be woken up if the No Sleep rodent triggers a floor rotation.

In this way, you've got two rats in identical conditions, exposed to the same environmental stimuli, except that one of them, randomly, has been cut off almost entirely from sleep.

SLEEP-
DETECTING
APPARATUS

Z

In the paper introducing this method for rats, the differences between the No Sleep and Control groups were pretty grim. The No Sleep rats did, in fact, die after anywhere from five to thirty-three days after the experiment began. They also didn't look good along the way, suffering from skin lesions, weakness, and swollen paws. The Control rats, according to the researchers, were "groomed, motorically active, and responsive to stimuli"—in other words, good to keep going.

So the disk-over-water method is one of the things that would be going through my head as I said "Yeah, probably" to the "Would you die if you never slept ever again?" question, though it still doesn't clear up the question of *what* about the sleep loss, specifically, caused the rats to die.

More recent work has illuminated this a bit. Researchers in 2020 reported that sleep-deprived fruit flies experience a massive buildup of "reactive oxidative species" in their gut which gets worse and worse the longer they're kept awake (until they eventually die). The same buildup was seen in mice kept on shorter sleep deprivation protocols. This by itself is interesting, but the kicker is that if they gave the fruit flies antioxidants (think: blueberries) during the sleep deprivation, *they no longer died.*

So despite the very obvious brain problems (hallucinations, irritability, being bad at basic things) you get when you stay up for long periods of time, the organ responsible for doing you in on a no-sleep schedule could very well be your gut.

Sure, sure, sure, in flies and mice. But what about humans? There, my mind would go to fatal familial insomnia, or FFI. FFI is a genetic condition that affects a very, very small number of people (one to two per million) and is inherited (hence, "familial"). It's a prion disease, like mad cow disease, and it's marked by insomnia that gets worse and worse until you just stop sleeping altogether and eventually die ("fatal").

The thing is, there's more to FFI than just not sleeping. The sleep loss is a symptom of it, sure, but you've also got *a prion disease* that is causing proteins in your brain to be *folded wrong*. That's going to lead to neurodegeneration, and it's that neurodegeneration that's thought to kill you if you're a member of one of the seventy families worldwide unfortunate enough to carry this gene.

The fact that FFI is so much more than just sleep deprivation, coupled with the fact that some rats died from the disk-over-water experiment on timescales much shorter than what we know humans can withstand, sleep loss–wise, convinces me that neither is a particularly great proof point for what it would take to kill a human with no sleep.

If, one day, I awoke strapped to a massive, human-sized disk-over-water setup invented by a Jigsaw-style villain determined to punish humanity for our sins against lab rodents, I'd think *fair enough* and *this is probably going to get me.* But I'd honestly be very motivated by that one rat that lasted thirty-three days in the original disk-over-water protocol. *I'm coming for your record, rat,* I'd tell myself as I hallucinated all the great luminaries of sleep medicine yore taking notes on my deteriorating condition. *Victory,* I'd think on day—I don't know, thirty-six, aspirationally?—when I finally fell off the platform and could no longer keep my head above the six inches of pool water.

I cannot stress enough how thoroughly and completely I think you should not cut your sleep duration short like this—not to the extremes of Z, nor Randy Gardner, nor the 1968 participants. The moral of their stories is not "Sleep (*smug, knowing smirk*), who needs it?"

But the way bad sleep is going to get you the vast majority of the time is through the steady accumulation over years of too-short, grooveless sleep. If you're lying awake at night stressing about a bad night's sleep, know that your basic biological operating system is, in some sense, remarkably robust to short-term acute sleep distress. Sleep loss is the kind of thing that kills you slowly.

If it kills you fast, it's because you fell asleep while driving.

CHRONIC SLEEP RESTRICTION

Let's move away from the (mostly lab-restricted) land of no sleep and into the (imminently more likely in your day-to-day life) land of *restricted* sleep. We have an idea of what happens if you don't sleep at all. What happens if you sleep two hours a night? What about four? Six? These questions map out a continuum, which goes from "Ha, surely only a weirdo who didn't think humans needed to sleep would try a schedule like this!" to "It is plausible that your air traffic controller has been on this exact sleep schedule."

In a highly impactful 2003 paper looking at *chronic sleep restriction* of this kind, the researchers put participants on different amounts of nightly restricted sleep and gave them a series of performance tests to see how their ability to operate broke down as time went on.

One of the performance tests they gave them was the psychomotor vigilance test, or PVT. Here's how that test works: You have people looking at a blank screen. Suddenly, randomly, a number appears and starts counting up elapsed milliseconds. They have to click it or tap it as quickly as possible. How long it takes you to respond is your reaction time. If you can't tap it within a set period of time (say, five hundred milliseconds), it counts as a "lapse."

Lapses are one of the most illuminating outputs of the PVT because they reflect a sort of "you didn't even notice there was a number" type of mental failure. A lapse usually doesn't occur because someone was just a hair late on the tap, coming in at exactly 501 milliseconds. They usually miss it happening all together, almost as if part of their brain was turned off. Not an ideal thing to have happen, especially if you're doing something like driving when the lapse occurs.

When you're giving someone a PVT test of this kind, you usually make it pretty long—say, ten minutes. This is by design. A very sleepy person can still muster up the focus to do decently well at a one-minute test, but when you intentionally bore them by having them look at nothing but a screen with some occasional numbers on it for an agonizingly long time,[21] they start to lapse more and more.

The researchers in this study had one group sleeping not at all, one group getting a four-hour "sleep opportunity," one group getting six hours, and one group getting eight. (The fact that it was a *sleep opportunity* and not true sleep is a subtle

---

21 *Longer than an unskippable YouTube ad. Longer than a six-minute comedy sketch from 2005 that you thought was great when it came out, but now, as you're showing it to your friends two decades later, you realize really should have been edited down a bit more.*

but important distinction here—a person in the six-hour group was getting no more than six hours but may have gotten less).

The daily PVT performance on the different sleep schedules looked like this over the course of the study:

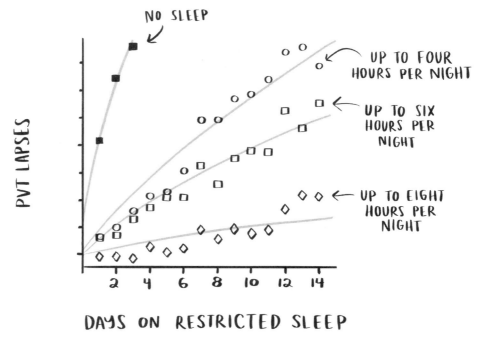

FIGURE 1A from Van Dongen et al., 2023

That is, everybody got worse and worse as the study went on, and how fast you got worse depended on how short your sleep window was. No sleep is basically a rocket to Lapse Moon.

Even the eight-hour sleep opportunity group showed signs of getting worse and worse, albeit much slower, which to me suggests they either needed more sleep than eight hours, weren't sleeping the full eight hours they had available to them, or were just getting fed up with the awful boring tests they were having to do every day.[22]

---

22 *Evidence that they weren't getting enough sleep within the eight-hour sleep opportunity can be found in a different paper, where performance deficits didn't accrue on a nine-hour sleep opportunity (but who knows; maybe those people just had a higher tolerance for boredom).*

There's another graph in this paper that captures the *subjective* self-reported sleepiness of the participants and tells us something very important:

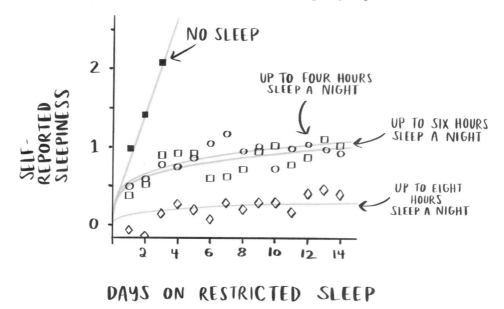

FIGURE 1B from Van Dongen et al., 2023

People in the no-sleep group: yeah, they knew they were tired and said so. People in the four- and six-hour groups, however, felt about as sleepy as they did on day seven as they did on day fourteen. *But both groups were twice as bad at the reaction time test on day fourteen as they were on day seven.* They were twice as likely to have a lapse as they were a week prior, but they felt basically the same.

In other words, our perception of how sleepy we are on restricted sleep schedules like these—and therefore how bad we are at things we're trying to do when we're on them—doesn't grow at the same rate as our actual impairment does. You're too sleepy to accurately gauge how sleepy you are. You're like Z in 1933, insisting to the researchers that *hallucinating* and *getting lost* is just normal for you while they side-eye each other over their clipboards.

Often, when I'm asked by people if I think they have a sleep problem, my first question is, "Do you think you do?" Because if you're feeling generally good about your sleep, and the only thing you're stressed out by is your sleep watch's deep sleep percentage, my response is, "You're probably fine." But just to make sure, I then ask, "Do your friends and family think there's something wrong with your sleep?"

Because it's possible to be so used to being tired that you no longer realize how tired you are (and how much it's affecting you).

FUNCTIONALLY DRUNK: SLEEPINESS AS BAC

You can give the PVT to people who aren't sleepy, of course. Drunk people also have been given reaction time tests like this. They do badly on them. This lets us come up with ways of going from "days on restricted sleep" to "functional blood alcohol content." It's not a great look for any of us who have driven while sleepy:

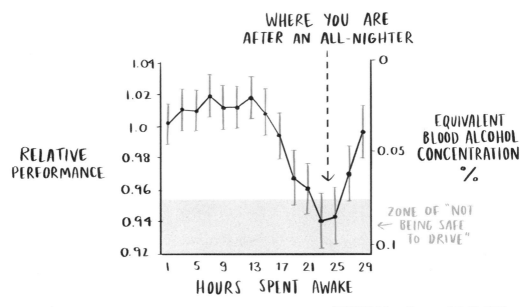

FIGURE 2 from Dawson & Reid, 1997

I think one of the reasons it's socially acceptable to drive while sleepy in a way that it's not socially acceptable to drive while drunk is that drinking is fun-coded, while not-sleeping is virtue-coded. People act as though not sleeping is a sign of Puritan righteousness (*you were working so hard, you didn't even have time to sleep*), and they forgive the act of fatigued driving because it's wrapped in a protective bubble of noble work ethic. The idea that it's a sign of either doing too much or managing your time poorly doesn't come into the picture. So two people get on the road, equally dangerous, and one of them is a worthless good-for-nothing (drunk person) while the other is an upstanding laborer (tired person). And either one of them could kill somebody.

Let me make the obvious defensive point here: some people's work schedules really do make it impossible to sleep. If you work twenty-four hours at a hospital and get on the road right after, you're doing that because your job demands it and not because you can't plan ahead. Or if you're given ten hours off between shifts and told to sleep the whole time, good luck with that! Odds are, you'll simply not be able to—first, because humans aren't machines and need time to get home and wind down, and second, because it's completely possible that your body's internal clock will be actively sending "be awake" signals that entire time.

If this is you, then of course the system is to blame for putting you on the road while dangerously sleepy. It's still a public safety issue. My stance is that employers should take responsibility: They keep you up too long? They get you an Uber.

OTHER BAD THINGS THAT HAPPEN TO YOU
IF YOU HAVE A SHORT NIGHT OF SLEEP

It's not just that you get bad at reaction time tests—short sleep is linked to other negative health outcomes as well.

Putting people on restricted sleep schedules of four hours per night impairs their glucose tolerance, ups their cravings for carbs, and boosts their markers of inflammation. People also get more sensitive to pain on these restricted sleep schedules.

In one of my favorite studies, the researchers brought in 153 healthy volunteers, asked them to self-report on their sleep habits over the last two weeks, and then *gave them nasal drops full of germs.* Then they stuck them in quarantine to see if they got sick.

The participants who self-reported less than seven hours of sleep in the weeks leading up to the germ exposure were about three times more likely to get a cold than those with eight or more hours (and people with low *sleep efficiency*—highly interrupted sleep—were five and a half times more likely to get sick than those with high sleep efficiency, or not very interrupted sleep). How well-rested they felt in the lead-up was not a good predictor of the likelihood of getting sick.

Ah, you might be thinking. *Thank you, science, for giving us such a nice, unambiguous picture of how more sleep is good for you and less sleep is bad for you. Truly there is nothing else to think about here.*

To which science would laugh in your face. Of course it's more complicated than that.

THE U-SHAPED CURVE

We can't talk about sleep duration and how it interacts with other aspects of health without talking about the letter *U*.

There's this inconvenient thing that happens when you look at sleep duration vs. "bad thing" in nearly all large studies that have that kind of data: the relationship isn't usually a straight line.

Sure, if you look at sleep duration vs. bad thing from, say, four hours a night to seven hours a night of sleep on average, you'll usually see the trend you expect, with *less* sleep duration associated with *more* of the bad thing.

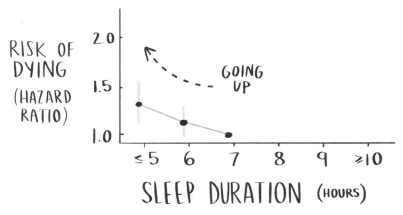

FIGURE 2 from Wang et al., 2020

But what about looking from seven hours a night to ten hours a night? Often, you see the direction of the line reverse, with *more* sleep associated with *more* of the bad thing.

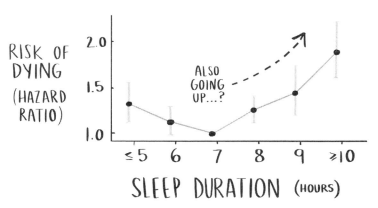

Putting these two trends together, you end up with a U-shaped curve. U-shaped curves happen all the time in studies looking at sleep duration:

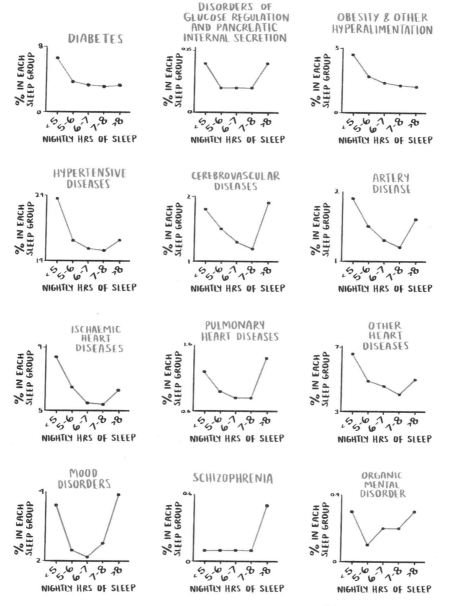

FIGURE 2 from Zhu et al., 2021

If more sleep duration is good, why would it ever be linked with something bad? The typical answer is one of correlation and causation. For people sleeping fewer than six hours a night, the short sleep duration is thought to be *causing* the bad thing. But for people getting nine hours or more of sleep a night, the fact that they're sleeping so long is simply *correlated* with the excess of sleep. Something else is making them very sick, and that sickness is also causing them to both sleep for a long time and have a higher risk of the bad thing happening. Or, in other words:

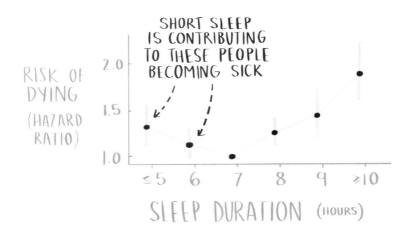

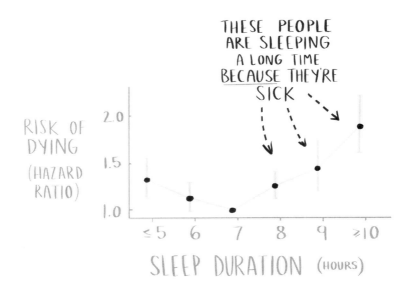

I think this is usually the right line of thinking. In general, though, distinguishing causation and correlation in large, retrospective studies is very hard. It's also very hard to rule out that correlation isn't a factor for the short sleepers as well. After all, you could be so sick that it's hard for you to fall asleep and stay asleep, making it so your net sleep duration is pretty short. The same sickness could increase your risk of bad outcomes as well.

And while it's very easy to disturb people's sleep in the short term in controlled laboratory environments—which is how we know that chronic sleep restriction makes you terrible at reacting/memorizing/learning and terrible at knowing how terrible you are at it—you're not going to have participants develop chronic kidney disease over the course of a three-day, in-lab sleep restriction protocol. You can't prospectively and randomly assign someone to a lifetime of short sleeping[23] just to know, definitively, if their later problems in life were *caused* by the short sleep durations.

What my heart craves, given the popular focus on sleep duration as *the* definition of sleep health, are more results like the psychomotor vigilance test reaction time assessments, where the longer you go with too-short sleep, the worse things get, in essentially a straight line. There, the result is so clear! It's bad to get four hours of sleep a night! Look how much worse you are with each passing day!

We don't usually get that with retrospective sleep duration data. Of course, that doesn't mean duration doesn't matter—the entire U-shape could be driven by the types of confounding correlation/causation influences I mentioned. But it does make you want to know if there are *other* dimensions of sleep health that could help to paint a clearer picture of what's going on under the hood. We'll come back to this idea in the next chapter.

OH GOD, AM I SLEEPING TOO MUCH?

There are two dangers in talking about the U-shaped curve for sleep duration. One is that people might see it and think, "Ha! Short sleep is just as bad as long sleep. [*Unspoken logical leap here.*] Short sleeping must not be that bad." The other danger is that people might start stressing out that they're sleeping too much.

To which I say, whoa there. We've got nice, tidy *causal* data for why sleeping too little can cause bad outcomes. We've literally kept people on restricted sleep protocols and seen their endocrine and metabolic machinery start to go on the fritz.

---

23 *As in, "Hello, six-year old child. Please, as part of participating in this research study, never sleep more than six hours a night for the rest of your life."*

What happens if you put people on a *sleep extension* protocol, where you invite them to sleep as much as they want for days on end? Do they *over*sleep, the way a person (me) might overeat at an all-you-can-eat Indian lunch buffet?

The answer, at least for healthy folks, appears to be a straightforward "no." This conclusion is based on a 1993 paper that I'll refer to as the Long Dark study, in which the authors first gave participants eight hours of darkness to sleep in per day for a week,[24] and *then* cranked the darkness knob up to fourteen hours of darkness per day.

But this time, they didn't keep people on the extended night schedule for a week. They did it for twenty-eight days. They put people in fourteen hours of darkness per day for a month.

Here's what happened to their participants' sleep:

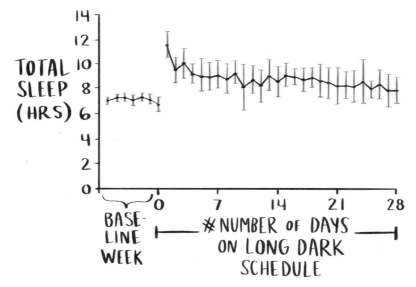

FIGURE 10 from Wehr et al., 1993

On the first night, people slept for a *really long time*. Their sleep period—the difference between when they fell asleep and woke up—was thirteen and a half hours, and they were properly asleep for almost twelve of those (and awake in bed for the remainder).

---

24 *Eight hours is a pretty short duration for a "night" as defined by the sun, but it's also pretty typical in terms of how long we give ourselves to sleep in modern life.*

Then they slept 10-ish hours, then 9-ish, until they leveled out around 8.6 hours or so, with levels staying roughly constant for the remaining weeks of the study. The general pattern was one of *exponential decay*, the same kind of faster-then-slower drop-off you see when hot food cools its temperature by sitting out on a counter.

Note that the sleep period (from when they first fell asleep to when they woke up) was longer than their sleep time because of nighttime awakenings—by two and a half hours or so. The nighttime awakening patterns looked like this:

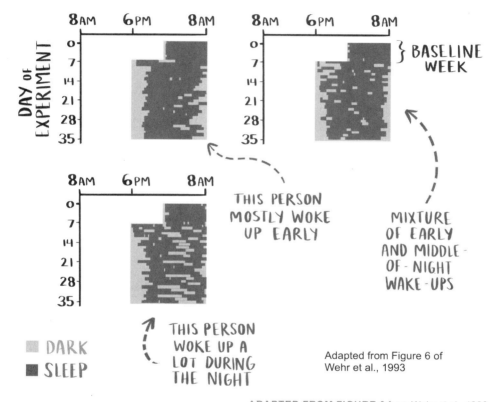

THIS PERSON MOSTLY WOKE UP EARLY

MIXTURE OF EARLY AND MIDDLE-OF-NIGHT WAKE-UPS

} BASELINE WEEK

DARK
SLEEP

THIS PERSON WOKE UP A LOT DURING THE NIGHT

Adapted from Figure 6 of Wehr et al., 1993

ADAPTED FROM FIGURE 6 from Wehr et al., 1993

Middle-of-night awakenings, some of them quite long, were perfectly common for these people. Yet, with one big exception, they had more energy and were happier after going on the Long Dark protocol. They felt significantly less daytime fatigue by the end of the twenty-eight day period, and two of the subjects had an extremely hard time adjusting back to regular life after the study ended—they'd gotten a taste of what it was like to not feel tired all the time. (The one exception was a subject with a family history of affective disorder, who became so depressed after five days of long nights that the researchers took him out of the study over concerns for safety.)

There was significant interindividual variability in terms of how long they slept during the long nights, with sleeps taken in isolation ranging from 5.2 to 11 hours, but nobody consistently slept longer than 10 hours after the initial transition period of increased sleep passed. The sleep banquet was on the table, but none of the subjects overindulged. As the days went on, they spent less of their sleep opportunity asleep and more time in the dark but conscious, either through waking up in the middle of the night, taking longer to fall asleep, or waking up early.

This heavily implies that you can't sleep too long—your body puts a natural stop to things on its own. That said, if you *are* consistently sleeping longer than ten hours, it's a sign that you should go talk to a doctor. Not because you're some greedy sleep glutton that lacks Protestant work ethic—because something else might be going on with your health.

SLEEP DEBT

A great thing about the Long Dark study is that it gives us a very tidy way of thinking about *sleep debt*. Sleep debt, in this framing, is the excess sleep from the first week or or so of that experiment—the area under the curve that's still above the steady-state level.

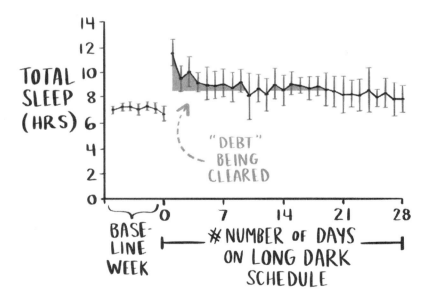

This period of elevated sleep that occurs before the sleep-per-night curve flattens out can in turn be thought of as a wash-out period for your sleep debt. The fact that the participants were on a seemingly very normal sleep opportunity schedule (eight hours of darkness) the week before the long nights started suggests that we may all be walking around with an accumulated sleep debt all the time. We're functioning, but we're lugging around a sleep need burden—one that could be cleared away if we just took two weeks or so and hung out in the dark for fourteen hours each day.

There's a notion of "sleep banking," where if you know you've got a late night of work coming up, you can take a nap ahead of time to help you handle the sleep loss better. From the perspective of sleep debt described here, banking of this kind isn't so much giving you an advance on sleep as it is relieving a chunk of the debt you always carry with you. It's releasing a bit of built-up *sleep pressure*.

This doesn't mean that the nap's not a good idea or that you won't do better for it! But it does underscore an important point: If a bunch of people who've been sleeping eight hours a night can suddenly sleep twelve hours when given the opportunity, then it's reasonable to assume a lot of us live our lives running a sleep deficit. And the fact that these same people were still sleeping ten-ish hours after two days on the Long Dark schedule suggests our current two-day weekend isn't nearly long enough to clear that deficit away.

PEOPLE WHO LIVE IN PRE-INDUSTRIAL SOCIETIES DON'T (ALWAYS) SLEEP EIGHT HOURS

Ah, modern society. If only you weren't running us all so ragged, we'd all be getting the long, luscious hours of sleep we so desperately need. If we could just sleep the way people sleep in societies without phones, we'd get eight hours a night for sure.

Except that's not what happens. In one study looking at sleep in pre-industrial societies, the authors found subjects sleeping an average of 5.7 to 7.1 hours per night, on the lower side of what most modern societies report. The people in that study went to sleep about three hours after the sun went down, slept an extra hour in the winter vs. the summer, and usually woke up before the sun came up. When the authors asked them if they had chronic sleep problems, only a tiny fraction (1.5 to 2.5%) said they did—way, way fewer people than you'd get if you asked the average American today the same question.

You can look at this and say, "Ha! Modern life *hasn't* affected our sleep after all!" which I think is pretty definitely wrong. For starters, the people in that study spent *much* more of their time in the dark after the sun went down, which affects your sleep by way of your circadian rhythms (more on this soon). Other studies have explicitly compared societies with electricity, where darkness in the evening is optional, to similar societies without electricity and seen that the addition of electricity delays sleep by half an hour to an hour.[25] Still, people in pre-industrial societies getting fewer than eight hours of sleep per night, especially in the summer, is a fairly consistent finding.

So how do we reconcile this with the findings of the Long Dark experiment and the chronic sleep restriction experiments? How much sleep do we actually need?

THE CONTRADICTIONS OF SLEEP

Let's dig in on the sense of cognitive friction you might be feeling.

Sleep is critical enough that we spend about one-third of our life doing it—*but* we have no data from humans showing that if you don't do it, you die.

Chronic sleep restriction makes you progressively worse at things like learning, immune response, and metabolic regulation—*but* people in pre-industrial societies don't necessarily sleep longer than people in the industrial world. Just because something happens in an "pre-industrial" context doesn't mean it's healthier or natural—*but,* at the very least, these people seem to feel less bad about their sleep than we do.

If you only give people an eight-hour opportunity to sleep, it's not a long enough window to avoid seeing performance drop-offs and a buildup of sleep debt—*but* if you give them a very long sleep opportunity, they won't sleep the whole time after their sleep debt washes away (as in the later days in the Long Dark experiment). In fact, they they don't fall asleep very quickly and wake up in the middle of the night the same way a *chronic insomniac* might (minus the "insufficient sleep" and "feeling bad" part).

And, extra unfairly, the fact that the people in the Long Dark experiment *still* sometimes had short nights of sleep (despite reporting overall that they felt great) suggests that a bad night of sleep in isolation isn't too much to worry about—*but* chronic sleep restriction studies also tell us that we can't reliably judge our own levels of impairment when we're sleepy.

---

25 *With sleep durations in one study going from 7.0 hours per night without electricity to 6.3 hours with electricity during the summer, and from 8.5 to 7.5 hours during winter months.*

And if you want to sleep more, one of the worst things to do is stress out about it. So calm down and don't freak out about the fact that you may be slowly killing yourself without realizing it.

What gives?

I have three answers to this question. The first is that not every sleep study measures sleep in the exact same way (nor are the sleeping conditions always equivalent), so the numbers that come out of sleep studies might not be comparable. This is a fiddly, box-checking answer, but it's one that does actually matter if we're trying to pin down a line separating good sleep durations from bad.

The second is that daily sleep need *isn't* some static, one-size-fits-all number for all people, or even the same person from day-to-day.

The third answer, and the one I'm most passionate about, is that sleep health encompasses much more than simply sleep duration. If we want to understand how someone's sleep behaviors are shaping their health, looking at duration of sleep alone is not enough.

I'll walk you through all three of these in what's left of this chapter.

PEOPLE ARE MEASURING SLEEP IN COMPLETELY DIFFERENT WAYS

Say you see a headline that says, "[Some specific number of hours of sleep] is associated with [some outcome, good or bad]." Probably the way they came up with that specific number of hours was either 1) asking people how long they sleep; 2) putting a watch on them to track their sleep; or 3) gluing electrodes to their head to watch their brain waves and then having a person look at the brain waves and decide, based on a host of factors, if they were asleep or not (this is called *polysomnography).*

All of these methods are wrong in their own special ways.

Asking somebody how long they slept is wrong because people aren't great at remembering things like that. Even if you ask them to write down how they slept right when they wake up, they tend to either over- or underestimate their total duration, with the direction of the bias tending to depend on how long their actual sleep was.

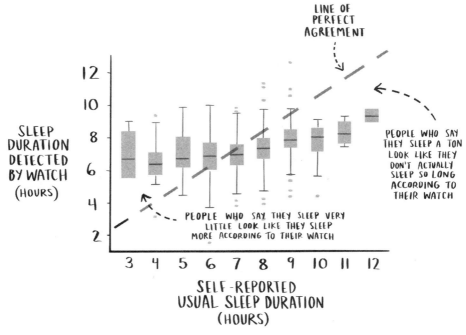

SLEEP DURATION DETECTED BY WATCH (HOURS)

12 10 8 6 4 2

LINE OF PERFECT AGREEMENT

PEOPLE WHO SAY THEY SLEEP A TON LOOK LIKE THEY DON'T ACTUALLY SLEEP SO LONG ACCORDING TO THEIR WATCH

PEOPLE WHO SAY THEY SLEEP VERY LITTLE LOOK LIKE THEY SLEEP MORE ACCORDING TO THEIR WATCH

3 4 5 6 7 8 9 10 11 12

SELF-REPORTED USUAL SLEEP DURATION (HOURS)

FIGURE 1 from Lauderdale et al., 2014

People also forget awakenings during the night all the time—both the many short ones that occur naturally, usually lasting fewer than thirty seconds, as well as the longer ones, where they're "awake" for minutes at a time. It can't be *that* surprising that self-reporting isn't a good measure of sleep: you were literally just asleep. It's like shaking a student awake in class and asking them what part of your lecture was so boring that they passed out during it. *How should they know?*

Putting a watch on someone (also called *actigraphy*) is wrong because your wrist is not your brain, which is where most of the action is when it comes to defining sleep. We've gotten better and better at extracting sleep information from wrist sensors, but there's still information loss in the journey from the skull to the end of your arm. So we can take high-resolution motion signals and heart rate signals at the wrist and attempt to infer from these what's going on in your brain, but the current standard is far from perfect.

WIDE-AWAKE BUT NOT MOVING

In fact, a lot of it boils down to, "Are you moving? You're awake. Haven't moved in a while? Probably asleep." Most of the sleep studies you read aren't using cutting-edge machine learning to determine your sleep state; they're using a rule like the "Are you moving? Then you're awake" test.

A great thing about approaches like these is that everyone in the research field uses them, which makes results comparable across research groups and studies. They even outperform some modern wearables from companies with multibillion-dollar valuations, in terms of how well they detect wake. A bad thing about them is that they still only correctly identify 40% of the wake you experience at night as being wake.[26] The reason is very, very simple: if you're awake but still for long enough, they think you're asleep. Your wrist is not your brain.

Polysomnography (PSG), the gold-standard measure of what sleep even *is*, is the most accurate way we have of knowing how long somebody slept, but it has its own problems. For starters, people tend to be weirded out by having a person glue electrodes to their head and watch them sleep, which leads to something called the "first night effect"—the sleep you sleep when you start nightly PSG is not representative of a normal sleep night for you (you usually sleep worse because it's all so strange). And since PSG is very expensive to run, rarely do you have more than one night of PSG for a person.

A more metaphysical answer to why PSG is still lacking as a way of capturing an individual's real-world sleep is that sleep isn't quite the categorical construct we act like it is. You're awake, we're told, and then you cycle through different stages of sleep—rapid eye movement (REM) sleep and non-REM (NREM) sleep—over the course of the night, with every tiny chunk of time mapping cleanly to a specific sleep stage.

But if you watch a human sleep lab technician score a night of sleep, it's not always so simple. There might be eye motions indicative of REM, but brain waves indicative of NREM. It can be ambiguous what stage of sleep you're in, or if you're even asleep at all. While more and more of this work is being shifted to AI, at present, it's something of an art and a science to decide what stage of sleep a person is in. And when AI takes over sleep scoring, if it's trained on human-labeled PSG data, some of that human artistic discretion will be inherently baked into the model.

A better way of thinking about sleep, then, isn't that we cycle cleanly through the primary colors of wake, NREM, and REM, but rather that the state of our brain

---

26 *Most modern consumer wearables are better than this, but not all of them! And their algorithms are changing all the time, so it's not easy to know if a device that was more accurate six months ago is still better.*

while we sleep is a gradient—a blur of all the different stages. Describing a person, at one moment in time, as being in 30% NREM sleep, 65% REM sleep, and 5% wake, for instance, would probably be a better expression of their true sleep stage than rounding up to 100% REM for that moment in time.

This isn't super relevant to the point I'm trying to make here, though, which is that there are three primary dialects being used when we talk about sleep, and they all have their quirks. Self-reported sleep will be skewed by self-report biases, actigraphy-based sleep will underestimate wake during the night, and PSG sleep will rarely be collected for multiple nights in a row (due to collection costs) and may be influenced by the strangeness of how invasive it all is.

There's a correction term to apply to any number you see bandied around in a headline or article that hinges on how sleep in that study got measured. And that doesn't even begin to take into account *other* factors that could influence sleep measurements across studies, like average participant age, room conditions, and the fact that people in pre-industrial societies aren't sleeping on super lush, comfy mattresses.

All of this, collectively, is a reason to de-emphasize hard numeric thresholds when making a binary decision about if your sleep is good or not. Given the lack of an exact translation between your Apple Watch's definition of sleep and a research paper's definition of sleep, focusing too much on how the paper's numbers and your Watch's numbers compare seems like an exercise in futility. Do you, independent of any number on any device, feel like you should probably be sleeping more? You should probably be sleeping more.

SLEEP NEED ISN'T ONE-SIZE-FITS-ALL

There are some people out there who truly seem to get by fine on very little sleep. You're probably not one of them—it's been estimated that this only shows up in about 0.004% of the population. That said, there are sizable interindividual differences in how people handle sleep loss, driven by factors like age and genetics. Moreover, we know your sleep changes qualitatively depending on how you used your brain that day—learning vs. non-learning, for example. I sleep more when I'm sick, and I sure as heck slept a lot more the day I ran a marathon I was massively unprepared for (vs. the night before, when I naively believed twenty-six wasn't *that* many miles).

In parsing the contradictions of sleep, natural variation in how long you, the individual, need to sleep today is always going to be part of the story. At the same time, it's not particularly actionable to shrug and say, "Well, everyone's different," especially when you, the individual, don't feel great about your sleep health and are looking for ways to improve. For that, we need to broaden our definition of what sleep health actually is, starting with acknowledging that . . .

SLEEP HEALTH IS NOT (JUST) SLEEP DURATION

Imagine you're looking at a city exactly from above, your gaze perfectly perpendicular to the ground. From your perch in the skies, you can see little people moving from building to building, like ants.

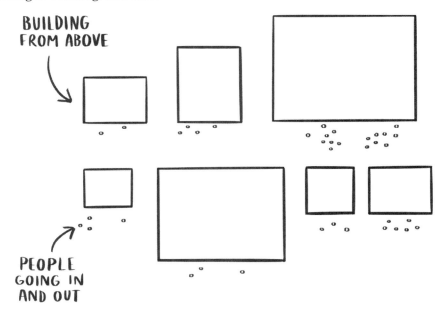

You can see that there are some buildings that a lot of people walk into and out of every day and some buildings that very few people frequent. The size of the building, as far as you can tell, seems somewhat correlated with how many people walk into that building, but it's not a clear trend. There are some buildings that seem big and have a lot of people going in and out of them, and some buildings that seem big and have very few people going in and out.

Of course, the problem is that, from your vantage point, you're not actually seeing the true size of the buildings you're looking at. You're seeing their footprint on the ground but not their height. Change the angle of your head, though, and suddenly the picture changes:

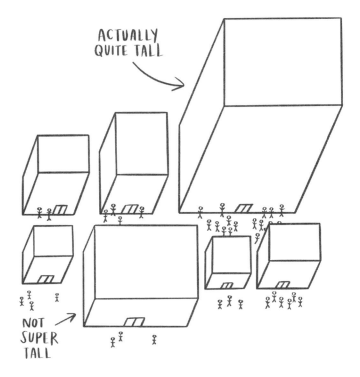

ACTUALLY
QUITE TALL

NOT
SUPER
TALL

All at once, what's going on becomes much clearer. There are buildings with large footprints on the ground that are *also* very tall, with huge capacities to hold people in them. There are also buildings with large footprints on the ground that are only one or two stories, with limited capacity to hold people within them. You didn't understand what was happening because you were flattening a 3D space into 2D. You were ignoring the other dimensions that matter.

This is what we're doing when we look only at sleep duration. Sleep duration is one dimension of sleep health, but it's by far not the only one. We can only really understand sleep health if we think of it as a multidimensional phenomenon.

This isn't something I invented. Plenty of others have thought about the multiple domains of sleep health, or "multidimensional sleep." Dan Buysse at the University of Pittsburgh has used the acronym RU-SATED as a way of highlighting key factors that matter for healthy sleep: **R**egularity of sleep, **S**atisfaction with sleep, **A**lertness during waking hours, **T**iming of sleep, **E**fficiency of sleep, and **D**uration of sleep. I'll discuss concepts like these more in the next chapter, including officially defining sleep efficiency.

But first I want to single out the sleep dimension I think is *most* undervalued by revisiting some data from earlier in this chapter. Many of the chronic sleep restriction studies I described, including the ones that tracked worsening performance on the psychomotor vigilance test, reported one data point per day. You can still have a juicy dataset with one data point a day, to be sure, but if you've ever pulled an all-nighter, you've probably experienced something like the following:

These are all *within-day* changes, and you would never pick up this up-and-down pattern if you were reducing each day to a single number.

Luckily, the researchers in the 1968 "Does sleep deprivation make you permanently psychotic?" study took plenty of within-day measurements. There, the researchers weren't just writing down the subjects' hallucinations and gripes during the study—they were recording their memory, mood, reaction time, and other physiological measures throughout.

You might expect pretty much of all of these variables to get worse and worse as the study went on, with self-reported sleepiness looking something like this:

It actually looked more like this:

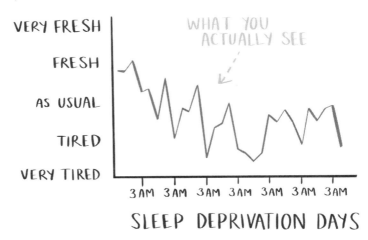

FIGURE 7 from Pasnau et al., 1968

The participants in 1968 would feel better, then worse, then better, then worse, with each better-then-worse transition happening on the timescale of about twenty-four-hours. Same for memory and reaction time tests. The ten-thousand-foot view of their performance was "bad and getting worse," but the *daily* trends had their ups and downs.

Yes. We return at last to *rhythms*. Even in research studies where sleep's natural patterns were expressly halted (by keeping people awake the whole time), there was still a steady thrum of a rhythm making itself known. The ticking of the body's circadian clock continuing, even as its connection to sleep itself was severed.

I'm going to spend the rest of this book arguing that the dimension of sleep health we most badly need to recognize and prioritize, society-wide, is this notion of sleep's *rhythmicity*—and the value of being in a sleep groove. Then, I'll go beyond that and claim that circadian rhythms are the ground beneath our feet, not just when we talk about sleep health, but in countless other aspects of health as well. Our bodies, from microscopic gene expression to macroscopic behavior, are intrinsically, inextricably rhythmic.

WHAT WERE WE TALKING ABOUT AGAIN?

A quick reality check. Say you're someone who sleeps one hour every day, at the exact same time, just for fun. You have a terribly low *sleep duration*, but your pattern of behavior is highly consistent. Your "rhythm" for sleep is the same every day. If you tell me you're not feeling great and ask me whether it's your sleep duration or your sleep rhythm that's driving your health problems, I'd smile sagely, lean back in my academic armchair, steeple my hands, and say, "YOUR SLEEP DURATION, DUMMY."

My hope is that this chapter has convinced you that sleep duration does matter, for a lot of things, even if the story is sometimes muddled by questions of correlation and causation, as happens with the U-shaped curve. In the case of sleeping extremely short durations each and every day, there's no ambiguity that it's bad and that rhythm won't rescue you.

But in the real world, we don't tend to see highly consistent hyper-short sleepers. What we see are people awash in a sea of conflicting forces: hunger for sleep, circadian signaling for sleep, social pressures to not sleep, and environmental factors undermining sleep. This leads to a wide range of expressed sleep behaviors, like the person who always struggles to fall asleep, or the person who falls asleep fast but wakes up frequently, or the person who can (possibly because of a massive unmet sleep need) sleep anytime around the clock.

Talking about this can quickly get complicated, which is why I'm now going to subject you to the terminology we've come up with to wrangle it.

READ THIS CHAPTER IF YOU WANT TO FALL ASLEEP

"AH," I SAY, WAVING YOU into the warm, candlelit room of this chapter. "Welcome." I gesture toward a comfy bed I've made up for you in the corner, soft and resplendent with blankets. On the opposite wall is a window, draped in thick curtains, that opens out to the night sky. Through a gap in the fabric, you can just barely make out a silent snowfall sheeting the earth.

"Rest, my friend," I say, patting the comforter, "and let me teach you some technical definitions from the field of sleep and circadian rhythms."

SLICING AND DICING A NIGHT OF SLEEP

Here are some of the numbers you might care about if you're trying to figure out if your sleep is good or bad:

1. **SLEEP DURATION:** how many hours you sleep, in total, over the course of the night

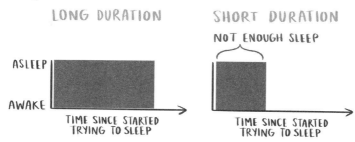

2. **SLEEP ONSET LATENCY (SOL):** how long it takes you to fall asleep

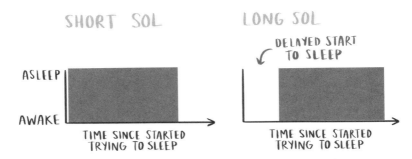

3. **WAKE AFTER SLEEP ONSET (WASO):** how much time you spend awake but trying to get back to sleep after you initially fall asleep

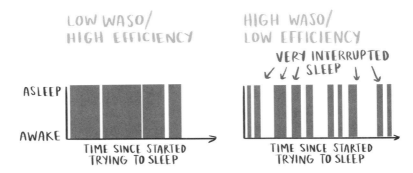

4. **SLEEP EFFICIENCY:** what percentage of the time you spend trying to be asleep that you actually spend asleep. High SOL and high WASO mean low sleep efficiency

I would consider all of these pretty standard sleep numbers. If you use a wearable, you might see them highlighted on your homepage or bundled into a "sleep score" that tells you how your sleep measures up.

Yet all of them are specific to a single night of sleep, in isolation, which is a bit like a doctor going in to measure your heartbeat with a stethoscope and stopping after hearing a single beat. Sleep, as I keep saying, is a rhythm, and a full snapshot of sleep health should care about this fact.

This means we need to think about numbers that capture this sense of rhythm. There are plenty of options.

Rhythms are things that repeat themselves, where one loop through a cycle takes about as long as the ones before and after.

A lot of rhythms in theoretical physics repeat themselves *exactly* every minute or *exactly* every 0.2 seconds, or *exactly* every seven hundred years. How long it takes for a rhythm to repeat itself is called its *period*. If you're taking a Physics 101 class, you're mostly going to encounter rhythms with periods that never change. Think "a spring on a frictionless surface, bobbing back and forth."

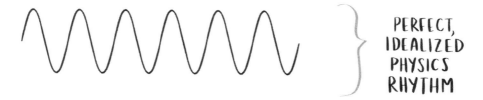

PERFECT, IDEALIZED PHYSICS RHYTHM

As long as nothing disturbs it, that spring will keep going back and forth forever, doing the same thing every time and taking the exact same amount of time each time it loops. Same thing is true of sine and cosine, if you remember those two from trigonometry class. They're rhythms, but they're really perfect rhythms. Unphysiologically perfect, if we're thinking about biology.

Biological rhythms tend to be stretchier and squishier than these idealized examples because inflexibility in biology can mean you just die. If your heart can't beat faster to help you run away from something, *not great*. If you can't hold your breath underwater, *not good*.

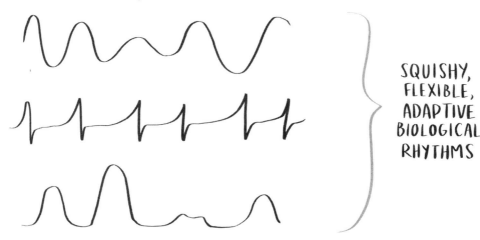

SQUISHY, FLEXIBLE, ADAPTIVE BIOLOGICAL RHYTHMS

So biological rhythms often cluster around one period (like your heartbeat's resting heart rate) but have the ability to speed up or slow down to meet your needs. Need to sprint? Heartbeat speeds up. Need to not breathe for a bit because you fell into a lake? Breathing slows way, way down for a bit.

Circadian rhythms, compared to breathing and heart rate, are pretty *inflexible*, period-wise. We couldn't adjust to life on life on planets with five-hour or thirty-hour days with our biology as it is right now, which may someday be a real, actual concern for humans as a species. We can still adjust to slightly shorter or slightly longer days, but we couldn't cut our period in half the way you can when you speed up you heart rate.

Said another way, while some biological rhythms can speed up or slow down a ton from one second to the next . . .

FIRST
SECOND

ONE SECOND
LATER

. . . a circadian rhythm is usually going to change pretty sluggishly from one day to the next:

FIRST
DAY

ONE DAY
LATER

A difference here is that while rhythms like heart rate are relatively *flexible* on period, they're relatively *inflexible* on amplitude. For circadian rhythms, it's the reverse.

Amplitude, in the case of the perfect rhythms from physics, is the difference between the highest and lowest point of the rhythm.

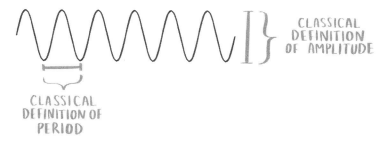

For biological rhythms, however, amplitude is much harder to define. You could say it's the difference between the highest and lowest point of the rhythm, or you could say it's the area under the curve, or you could say it's best reflected by the amount of time your rhythm spends above some threshold. We'll touch on amplitude again near the end of this chapter.

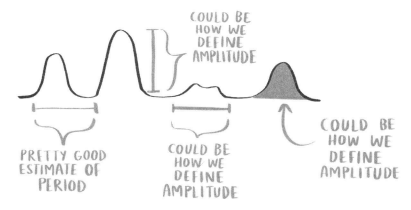

Still, the broader takeaway is that while you don't see speedy circadian rhythms, you can see very, very flat ones.

We can speculate why circadian rhythms tend to maintain period over amplitude: The concept of *jet lag* (body clock disruption caused by crossing time zones) isn't one that ancient humans would have been likely to encounter, so there wouldn't be a need for the circadian period to deviate very far from twenty-four hours. In a similar vein, there could be times when temporarily flattening your rhythms would be actively useful (e.g., a flatter *sleepiness* rhythm could make it so you're able to stay up more effectively during the night if you need to be continuously on the move).

Regardless of the reason, circadian rhythms stand out among the rhythms in our body for having a doggedly stable period of about twenty-four hours. This is even the source of the word "circadian": *circa diem* in Latin, which means "about a day."[27]

BREAKING DOWN A RHYTHM

When we talk about rhythms, amplitude and period are two important numbers. A third one is phase. Phase here means how far along through a single cycle you are at a given point in time. Biologically, it's usually best defined relative to something, like "hours after some event."

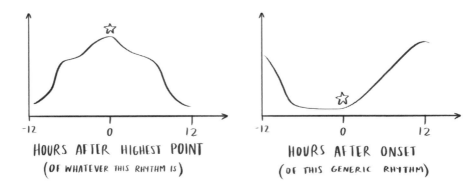

HOURS AFTER HIGHEST POINT
(OF WHATEVER THIS RHYTHM IS)

HOURS AFTER ONSET
(OF THIS GENERIC RHYTHM)

Circadian biology has a few favorite reference points for phase that come up over and over again. They are 1) the time your core body temperature reaches its lowest point, and 2) the time when the hormone melatonin would naturally start to rise in your body if you were in the dark. Truly, these are some of the nicest and simplest ones we have.

Core body temperature here is not the temperature you would get if you stuck a CVS thermometer under your tongue and looked at it over the course of the day. It's the temperature you would measure from somebody who had either swallowed a thermometer pill or had a rectal thermometer inserted. Also, this person is not allowed to move. Also, they're not allowed to sleep. You set them up in a room, monitor their temperature, and watch to see when the lowest point occurs. For the average person, this is around 3:00 or 4:00 in the morning, but the spread across individuals is enormous. Shift workers can hit their lowest core body temperature at any time in the day.

---

27 *By law, every talk, article, book, or poster about circadian rhythms must mention this fact at least once.*

Why not use a simpler way of measuring temperature to track circadian rhythms? It's because circadian rhythms control your temperature, *but a lot of other things affect your temperature as well.* If you get up and move around, that will heat up your body and cause your temperature to rise. If you get a fever, that'll affect your temperature too, as will the simple act of sleeping. But those changes are not your circadian rhythm at work. If anything, they're hiding your circadian rhythm. They're *masking* the true rhythm from view.

So by making a person somewhat miserable (no sleep, no movement, butt thermometer), you're able to measure a true circadian rhythm with relatively little noise or interference from other factors. And you'll often see *other* circadian rhythms defined relative to core body temperature's lowest point, also called *CBTmin*; i.e., peak athletic performance tends to happen ten hours before CBTmin. CBTmin is roughly when the participants in the 1968 sleep loss study felt their most fatigued each day.

Of course, measuring CBTmin as I've described it is a huge pain. It's also not compatible with realistic living conditions, which is one of the reasons why the field often uses *dim light melatonin onset* (DLMO) as an alternative way of telling biological time.

Melatonin is a hormone produced naturally by your body, and, like temperature, it shows a circadian rhythm. Think of it like your biological signal for nighttime. It isn't triggered in response to sleep, nor does it force you to immediately fall asleep, but it does help set the stage for sleep to happen in humans.

A natural rise and fall in melatonin might look like this:

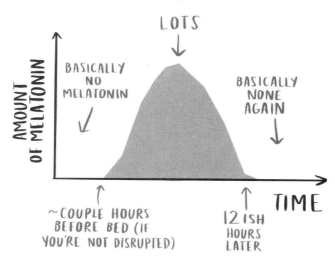

Unlike temperature, which is reliably above zero while you're still alive, melatonin truly vanishes and returns from your body daily. Put another way, your body is a natural pharmacy, and it's dosing you with melatonin once a day.

Or it tries to. Your body's natural production of melatonin won't happen if you're still in the light. Light exposure suppresses melatonin, probably because melatonin is supposed to tell the rest of your body that it's nighttime. If the lights are still on in the evening, our systems assume that the sun is still up and that melatonin production should hold off for a while. Light masks melatonin rhythms the same way movement could mask a temperature rhythm.

So when we measure melatonin in your body, we have to do it in the dark. If the lights are on, or even not dim enough, melatonin might refuse to come out, like a groundhog sticking to its den in a too-cold February. The way measuring DLMO often works is that someone comes into a lab, you turn out all the lights, and then you start having them spit into tubes once every half hour or so. Sometimes you do this for the six hours around their normal bedtime; sometimes you have them do it for an entire twenty-four hours. DLMO, then, is the time when the melatonin in your body starts rising and crosses a threshold that's different enough from zero to clearly indicate, "Okay, melatonin time is now."

I'd be remiss in saying that these two quantities, CBTmin and DLMO, are the only ways we have now, or will ever have, to track circadian phase. There are plenty of emerging methods in the research world, including techniques developed by my research collaborators and at my company, to predict circadian phase without needing spit or rectal thermometers. I could write a whole separate chapter on the shortcomings of core body temperature and dim light melatonin onset: they're collected in unrealistic conditions, they're still subject to finicky noise, they don't capture the full orchestra of rhythms in your body, etc.

Yet the reason we have DLMO and CBTmin now is because simpler methods, like using rules of thumb, or sleep timing, or looking at a person's activity patterns over a single day, or measuring temperature from a watch can all be totally, completely, 100% wrong at predicting true circadian time. The circadian signal gets "masked" in the measurements, and you end up twelve hours off of predicting their actual circadian time.

Here's an example of how a rule of thumb based on sleep timing can break down. In the average person, DLMO happens about two hours before bedtime, but like with CBTmin, the interindividual variation can be huge. Someone working night shifts might have their DLMO time at 3:00 PM despite sleeping that day from 7:45 AM to noon. Someone else on the *same exact work schedule* could have their DLMO at 3:00 AM.

And you don't have to work nights. Anyone who has experienced some kind of circadian disruption can have their DLMO, or any other marker of their circadian clock, flung to an unexpected time of the day. Traveling across time zones? Staying up late to get work done? No simple rule like "Melatonin production happens two-ish hours before bedtime" is going to work in all these cases. Your body clock is going to have its own idea of what time it is, and that time might be very unintuitive.

How? Let's answer that question by talking about how signals matter to your circadian clock.

TIMING SIGNALS

Zumba is a cardio dance class you can take at your local rec center. You follow an instructor through dance moves in time with peppy, upbeat music. When I'm doing Zumba at my local gym, I use lots of cues to figure out where the beat is: the auditory cue of the music, the auditory cue of people's feet hitting the ground, the visual cue of the instructor at the front of the room, the visual cues of people around me. Each one of these is a *timing signal*—they tell me where in the rhythm I'm supposed to be. If I see or hear that I'm a little bit behind the beat, I speed up. If I see or hear that I'm a little bit ahead of the beat, I slow down.

Light exposure is considered to be the dominant timing signal (or *zeitgeber*, which means "time giver" in German) for the central circadian clock in your brain, though exercise and activity behave in very similar ways to light. Taking melatonin as a pill, which you can buy in the U.S. over the counter, can also give signals about what time it is to your body's clock, although it tends to send the opposite message that light would send at that time.

These signals give us information about time, but they can't immediately "hard reset" our body's time. Instead, each tiny chunk of timing signal you get acts like a vote. Given enough votes telling the same story, our body's clock is gradually steered onto a certain schedule.

If I fell wildly out of sync while doing Zumba—say, for instance, I caught a glimpse of myself in the mirror and had a traumatic flashback to the dance routines I choreographed to the Spice Girls as a child—rarely would I be able to *instantly* snap back into sync. My arms and legs would be in the wrong place, for starters. There might also be conflicting timing signals, like the person in front of me not being exactly in sync with the instructor. I'd need time to look around the room and decide which signals to trust. Then I'd gradually integrate all the signals I was getting, weighting some (music, instructor) more than others, and ease back into the groove.

The timing of our clock markers isn't just determined by one part of one day, like when you were on shift last Tuesday or when you were sleeping on Friday. It's decided by the *cumulative* effects of all the timing signals you feed your body, minute by minute, continuously leading up to the present moment. This is how two people on the same work schedule can have wildly out-of-sync circadian clocks—by getting different signals during the hours they're not working.

More recent signals matter *more*: you're not likely to be still reeling from the aftereffects of "turning the bathroom light on overhead at 2:03 AM when you were eight years old" today. But if you stayed up super late or crossed time zones half a week ago? Yeah, you're probably still feeling it. Your body's still getting back into a groove.

ENTRAINING YOUR BODY'S RHYTHMS

This process of adapting to a rhythm, or getting into a groove, is often called *entrainment*. Entrainment, as a picture, looks like this:

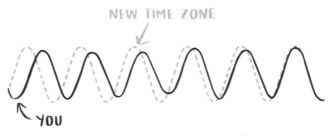

For your circadian clock, entrainment is the process of all your body's rhythms stretching and adapting themselves gradually to match a new repeating timing signal or schedule. This is what happens when you cross time zones or switch to a new wake-up time. When a new Zumba song starts and I get into its rhythm, that's me *entraining* to the new pattern of timing signals I'm getting. If the music is quiet, or I can't see the instructor, or everyone around me looks to be on a different beat, it'll be harder for me to find a signal to *entrain* to.

Here's another analogy for entrainment. Imagine you're walking on a sidewalk and you're doing that thing where you step first with your left foot and then with your right foot exactly once in each sidewalk square. [28]

---

28 *If you haven't done this, go outside and try it at once so you can understand my analogy.*

Probably, the gait you walk with when you're trying to hit exactly two steps per sidewalk square isn't *exactly* the gait you'd be walking if you were on a seamless stretch of concrete, without any squares. If your stride is naturally longer than the sidewalk square, you shorten it. If your natural stride is a little shorter than the sidewalk square, you lengthen it a bit. You're modulating your gait a little bit so that you can get that sweet, sweet two-steps-per-square in. You've *entrained* to this sidewalk pattern.

Now imagine you reach a driveway. There are no sidewalk squares for a stretch, and when you hit them again, your feet are out of sync with the pattern of squares—maybe you step first with your right foot instead of your left, or maybe you have to take three steps in a single square because of where your feet landed. But gradually, over the course of the next few squares, you get back into a left-right (new square) left-right (new square) pattern.

This period of time when you're stretching or shortening your gait to get back into your original groove is *entrainment* to the new pattern of sidewalk squares.

We can compare this transition period, when you're adapting your stride for the new sidewalk pattern, to the experience of crossing time zones during travel (a.k.a. jet lag). The time you spend walking on the square-free driveway is like the dim, liminal environment of a long flight, and your return to sidewalk squares that are misaligned with your current gait is like your arrival in a new time zone that's misaligned with your body's rhythms from before the flight. But give it a few days (sidewalk squares) and you'll re-entrain just fine.

EXCEPT WHEN YOU DON'T

Remember the point I made before, about how our circadian period can get a little longer or a little shorter, but otherwise is pretty tied to a 24(-ish)-hour day? In the sidewalk analogy, that's equivalent to "your legs can only stretch so much."

If you were trying to walk two steps per sidewalk square and I slowly started making the sidewalk squares longer and longer, like some kind of diabolical city planner, eventually your legs wouldn't be long enough to hit two steps per square, no matter how hard you tried. Similarly, if I made the sidewalk squares shorter and shorter, you'd eventually start tripping over yourself, to the point where you couldn't get into any sort of walking groove.

That's what happens to your body's circadian clock when the day gets more than a little shorter or longer than 24 hours. By 22- or 26-hour-long days, our circadian clock starts tripping over its metaphorical feet or surrendering to the inevitability of its stretchy limits.

So what happens if you can't or don't entrain to a schedule? If you're walking on the sidewalk and you're me, you just give up, ignoring the pattern of sidewalk squares and the joy that comes from getting two feet in each one. You walk at your natural stride, which, again, is probably not exactly the same size as the sidewalk squares.

In circadian parlance, you "free run." That means that you continue to move forward at a period that's slightly different from the 24-hour day. Parallel to the idea of having a natural stride, this is your natural, *intrinsic circadian period*. Most people cluster around 24.2-hour days, though it's possible to be either longer or shorter than that.

When you start free running (which can happen if you lose your eyesight and ability to process light, or just decide to stay in the dark all the time as a life choice), you'll probably only drift a little bit, e.g., by having your circadian markers like DLMO drift twelve minutes later every day. But one month of that is a six-hour shift in your body's internal time relative to where you started—the equivalent of going from the central U.S. to London. And if you keep free running, you'll eventually cycle back around to where you started, having gone through all "effective time zones" of the world, without even leaving your hometown.

When we lose the ability to entrain and free run at our natural period, the relationship between our bodies and the sun becomes like the relationship between the turn signals of two cars stalled at a stoplight, blinking in sync and out of sync and in sync again—the rhythmic equivalent of boats passing in the night.

WHY WE DON'T FREE RUN ALL THE TIME

Timing signals tell our circadian clocks what time it is (great!). We can be influenced by timing signals that happened weeks ago (okay!). Sometimes timing signals send conflicting messages, which can confuse our sense of time (ah). If we lose our sense of time, we become detached from the 24-hour day and start drifting forward under our own definition of what a day is (oh no).

With so many signals feeding into how our body determines biological time and the potential for them to send wrong, contradictory information, why don't we free run more often?

There's good news: in some sense, the physics of the circadian clock set it up for stability. Let's assume you're a person entrained to a day schedule, getting a very consistent pattern of light and dark that pretty much matches the solar day.

Now assume that your clock gets slightly out of sync and is running a bit behind schedule when Monday morning rolls around. When the light hits your eyes through your window that AM, it's going to be arriving during something called the *phase advance region* of your body's circadian clock. For an average person, the phase advance region might start around 4:00 AM and end around the early afternoon. [29]

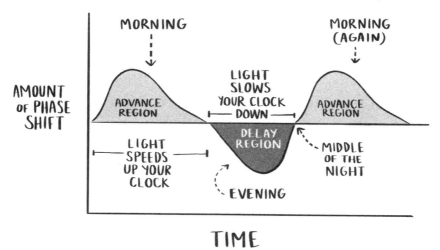

Light exposure during the phase advance part of your day speeds up your body's clock. It nudges the machinery of your clock to run a bit faster. How much it speeds up your clock depends on when you get it (you'll probably get more of an accelerating boost during the mid-morning vs. midday, for instance), but the net effect is a speed boost.

Your body clock in this example is running a bit behind schedule, so if you start getting light at your normal, consistent wake time, you end up getting more "speed up your clock" light than normal. This will have a *corrective* effect on your circadian rhythms, counteracting the fact that you started the day with your biological time running late.

Similarly, imagine your clock is running a little *ahead* of schedule and you're feeling sleepier than usual as the day winds down. Just like we have a phase advance region of our day, we also have a *phase delay region,* which tends to start in the afternoon and stretch into the early morning. Light exposure during the phase delay region slows your rhythms down.

You're running ahead of schedule in this example, so you want to spend more time in the phase delay region than you normally would to "correct" for that. But the

---

29 *But again! Everyone is different! This won't be the right time for you if you've recently disrupted your clock, work shifts, or have any other biological time apart from the "roughly average" one I picked for this example.*

good news is that, since you're already running early, simply keeping the lights on until your normal, consistent bedtime will get you that extra bit of slowdown. You just need to do what you'd normally do, and the fix will be baked in.

We can plot the extent to which different signals speed us up or slow us down in *phase response curves* (PRCs). They tell us how much a jolt of light exposure (or some other timing signal, like activity or taking melatonin as a pill) will speed up or slow down our body's rhythms at any point in time.

Different brightnesses of light and durations of light will have slightly different shapes of phase response curves, but they generally tell the same story of light speeding you up in the morning and slowing you down in the evening. Exercise's phase response curve tells a similar tale:

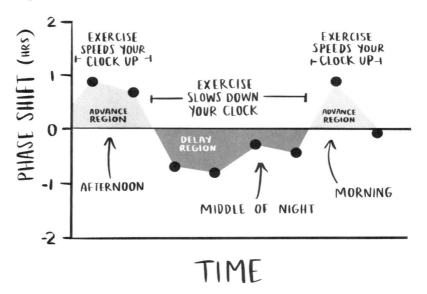

ADAPTED FROM FIGURE 4B from Youngstedt et al., 2019

These shapes reflect a sort of built-in protection for your circadian rhythms, like bumpers in a bowling lane. Get a bit off track? Stick to your normal schedule and you'll lock back on.

GOING OFF THE RAILS

In both of the examples I've given here, small disruptions to your body's rhythms are naturally corrected by sticking to a consistent lighting schedule.

There are two key phrases in the last sentence, the first of which is "small disruptions" and the second of which is "sticking to a consistent lighting schedule." The dark side of light exposure at night is that once you go beyond a spoonful, it actively *discourages* consistency. We live in a world where light is under our complete and terrible control, and consistency is optional. Choosing consistency, then, is like choosing to cut back on your sugar intake. Light at night in this analogy is like walking next to a Cinnabon at the mall.

When you get light at night, in the phase delay region part of your day, it delays your clock, which causes your "time when you'll feel sleepy" to get shifted later. And if you're not sleepy, you're more inclined to keep the lights on because you want to be awake and do things. So then light delays you even more.

This is a *positive* feedback loop with a *negative* end result: you end up much more delayed relative to where you would be if you didn't have the power to control your light exposure (for instance, if you were camping). This kind of phenomenon could be the reason night owls exist to the extent they do in modern society: if you're more sensitive to light (and some people are really, really sensitive to it), you are more susceptible to falling into the whirlpool trap of light at night delaying you, and delaying you, and delaying you some more. We'll come back to this in later chapters.

RHYTHMS CAN INTERFERE WITH EACH OTHER

There's another important property of rhythms, which is that they can boost each other or cancel each other out.

When two waves collide in sync, they can add to each other, making their amplitude bigger:

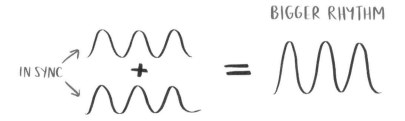

Or they can collide out of sync and cancel each other out, making their amplitude smaller:

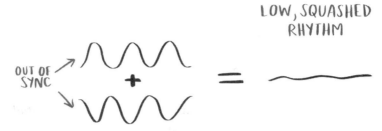

My preferred analogy for this is "being on a trampoline." If you've never trampolined before, it's a great way to have fun as a kid and then break your ankle in your thirties. [30]

If I'm sitting on a trampoline next to some kids bouncing, eyes closed so I don't have to see anyone break their ankle, I can still track what's going on if they're all bouncing in sync. *Ah,* I'll think, as their combined weight pushes the surface of the trampoline down. *They're at the bottom of the jump.* There's a clear sense of collective "up" and "down."

If they're jumping at all different times, however, I won't have a clear sense of rhythm. There will be lots of ripples in the surface but no obvious sense of "this is the group's in-air time" and "this is the group's return to earth."

I bring this up because the timekeeper in your brain is like a massive trampoline with tens of thousands of bouncing kids on it. Some of the kids pay attention to light exposure and jump faster when it's bright outside and slower when it's dark outside. Other kids are watching them and trying to time their bounces to sync up with everyone else's.

Swap in "neurons" for "kids" and "action potentials" for "bounces" and this is roughly how the *suprachiasmatic nucleus* (SCN), the central circadian pacemaker, behaves.

How synchronized the neurons are—how much they're firing together vs. fumbling around—is maybe the purest definition of *circadian amplitude* that we have. If all the neurons in the SCN are like "Yes! It's daytime!" they'll send a clearer, stronger signal to the rest of your body (high amplitude), whereas if half of them say it's daytime while the other half say it's nighttime, the signal will be muddled and low amplitude.

---

30 *Go outside and try it at once so you can understand my analogy.*

The only problem with measuring amplitude this way is that it requires you to look inside someone's brain, which is hard. Seemingly easier ways of assessing amplitude, like looking at how much melatonin a person naturally produces, have their own issues (some people are naturally high melatonin producers, for example, but that might not have anything to do with their circadian rhythms). Coming up with a way of experimentally measuring circadian amplitude is an area of active research, and until we're able to, amplitude as a concept is a bit like Bigfoot—we've got a blurry idea of it, but we haven't pinned it down.

Yet, just like with phase response curves, we can define *amplitude response curves*. Most math models of circadian rhythms produce amplitude response curves that look like this:

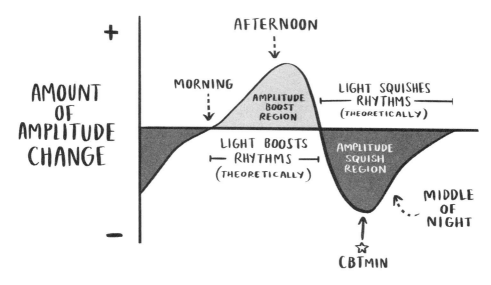

Light would be expected to boost your amplitude if you got it in the middle of your circadian day—say, twelve hours after CBTmin. Light would be expected to squish your amplitude if you got it right around CBTmin time. I think of this in the following way:

BRAIN (RECEIVING LIGHT IN THE MIDDLE OF ITS DAY): Ah, I am getting light when I expected to get light. This reinforces my belief that it is day. I am now more confident than ever that it is day and shall increase amplitude as a result.

BRAIN (RECEIVING LIGHT IN THE MIDDLE OF ITS NIGHT): WHAT IS HAPPENING. WHY IS THE SUN HERE.

I'm saying "expected" here because these results are, again, mostly theoretical. But there's suggestive experimental evidence that, sure, if you want to sleep earlier or later, get light in the phase advance or phase delay regions—and if you like the timing of your sleep and want to sleep *deeper*, get light in amplitude *boosting* parts of your day.

WHEN A GROOVE BREAKS DOWN

Phase, entrainment, amplitude: all these quantities we've talked about with rhythms can cause problems when they get thrown out of whack.

A phase problem could be feeling like you're doomed to always be a night owl while wishing you were a morning person. Or chronically waking up earlier than you'd prefer, then finding yourself totally exhausted by the time the early evening swings around.

An entrainment problem could be something environmental, like living in Antarctica in the winter, getting near-constant darkness, and never having an entraining signal to "lock" onto. Or you could be a person with non-24-hour sleep-wake disorder (non-24), a brutal condition, common among blind people, that causes you to free run all the time at your intrinsic period. This means being in sync with your family, friends, and loved ones, then falling out of sync with them, then in sync again, then out of sync again, for all time.[31]

Amplitude problems are less well studied, but they're something I'm particularly passionate about, having talked to a lot of shift workers over the years and heard lines like these over and over again:

- "I have no rhythms."
- "It's just a constant experience of bleh."
- "I get off shift, and, more than tired, I just feel . . . flat."

This, to me, *screams* amplitudes that are too squished. Which makes sense: when you simulate shift-work schedules through models that capture the physics of human circadian rhythms, you don't see nice, robust amplitudes.

You see amplitudes that look like they've been stepped on:

---

31 *It's awful.*

When your amplitude gets squashed, it messes with your sense of phase as well. After all, it's easy to find where you are in a high-amplitude rhythm, but it's not so simple when the amplitude gets low:

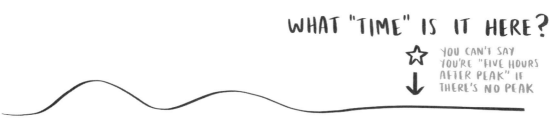

With no clear signal for time, and no clear signal for when to sleep as a result, a person with low amplitude could struggle to stay asleep for long periods of time. Other groups likely to suffer from low amplitude are older individuals, who tend to show across-the-board flattening of circadian rhythms, and people who spend most of their time indoors. [32]

Theoretically, these groups would benefit from aggressive light exposure during the daytime hours, especially in the midday (vs. only at, say, the very start of the morning), and aggressive light avoidance at night. Those actions should boost their circadian amplitude, increasing the likelihood that they'll sleep without interruption during the night. Plus, since we're talking about circadian rhythms, which aren't *just* sleep, the benefits of higher amplitude might not *just* be sleep either. [33]

There might be others who could benefit from lowering amplitude as well. People trying to cross time zones should have a much easier time doing so if their starting amplitude is low. And wake therapy, an intriguing treatment for depression, works by having people stay up all night in bright light, right at the time when light is most expected to squash their body's amplitude (and confuse the heck out of their SCN). It's possible that this antidepressive effect could be linked to amplitude, but we don't know.

So, there's a lot to explore in the concept of circadian amplitude, and when you look under the hood at what I mean when I say "sleep groove," the closest definition is "having a robust sense of day and night." Which is just amplitude.

---

32 *Even light that doesn't seem dark can still be very dim relative to the sun.*

33 *Take Mason et al, 2022, a paper looking at the non-sleep consequences of light exposure at night: "In healthy adults, one night of moderate (100 lx) light exposure during sleep increases nighttime heart rate, decreases heart rate variability (higher sympathovagal balance), and increases next-morning insulin resistance when compared to sleep in a dimly lit (<3 lx) environment."*

The bad news is that we don't have a good, non-brain-invasive way of measuring circadian amplitude. The good news is that it does, in theory, correlate with things that are much easier for us to measure.

Things like *sleep regularity*.

CIRCADIAN DISRUPTION AND SLEEP REGULARITY ARE DIFFERENT THINGS

Sleep regularity, at the highest level, captures how much your sleep timing looks the same from day to day. Going to bed and waking up at the same time every day makes for high sleep regularity. It's close to the concept of circadian disruption, but they're not the same thing.

Imagine you're a person who's very consistent with both your sleep habits and your light exposure habits. Every night at 9:30 PM you start dimming the lights in your home. At 11:30 PM, you go full lights-out and fall asleep shortly thereafter. At 7:00 AM the next morning, you wake up without an alarm.

Then, one random Tuesday, something stressful happens. You're not able to fall asleep at your normal time that night. At this point, the universe splits into two parallel timelines. In the first, you still dim the lights at 9:30 PM and turn all the lights off at 11:30 PM. You're awake, and you're out of bed, but you stay in the near-dark the whole time. Maybe you sit on the couch in the dark and listen to an audiobook, or you journal out your thoughts on your phone with dark mode turned on. You eventually fall asleep at 2:00 AM.

In the second scenario, you do essentially the same thing but without turning off the lights. You stay up and about in your normal living room, overhead lights on like it's still the middle of the day. You pace back and forth. You journal, but this time it's on your desktop computer. Dark mode might still be on, but with all the other lights still on in your environment, you're still getting a healthy dose of photons to the eyes. In this case, let's say you also happen to fall asleep at 2:00 AM.

In both scenarios, your sleep timing has shifted dramatically, from 11:30 PM to 2:00 AM. Your *sleep regularity*—a number that can be defined many different ways, but one that attempts, in all of them, to capture how much your sleep timing shifts from one day to the next—is going to be bad, at least from Monday to Tuesday.

But in the first timeline, your *circadian disruption*, or how far off your body's rhythms are from their ideal groove, will be lower than in the second. It won't be nothing (that delayed sleep, that dim mode on your phone will do *something* to your circadian clock), but it will be less than the disruption you'd experience in the second scenario, where on top of losing sleep by staying up late, you're implying to your body that the sun was up for an extra three hours that day.

OPTION A

OPTION B

YOU'RE UP LATE, BUT YOU DON'T THROW OFF YOUR CIRCADIAN RHYTHMS THAT MUCH

YOU'RE UP LATE, AND YOU THROW OFF THE GROOVE OF YOUR CIRCADIAN RHYTHMS

Circadian disruption and sleep regularity aren't wildly unrelated concepts, of course. In general, bad sleep regularity will mean disrupted circadian rhythms and vice versa. If you're staying up until 2:00 AM some nights and 8:00 PM others, you're not going to be sending your circadian clock a clear, unambiguous signal about when day is and when night is. Similarly, circadian rhythms that are stretched, squished, or otherwise weird will make it hard to fall asleep and wake up at the same time every day. Getting into a sleep groove requires a rhythm to groove *to*.

Yet they're different enough that it's worth repeating the moral of the story of the two timelines to really underscore it: you can more immediately and directly control the circadian disruption you experience than you can control your sleep. You're not powerless to control your sleep (we'll talk about strategies like cognitive behavioral therapy for insomnia and distracting yourself during rumination in the parts ahead), but reining in your circadian disruption can be as easy as flicking a light switch. That's not something you can do with sleep.

An important note: the regularity of your sleep is a completely different quantity from how long you sleep.[34] After all, let's go back to the example of a person sleeping exactly one hour every night, from midnight to 1:00 AM.[35] Your sleep duration would be ridiculously low here, but your regularity would be perfect.

So circadian disruption and sleep regularity are different but linked, while sleep duration and regularity are two distinct and uncoupled concepts. Since sleep duration and regularity are much easier to measure in the real world than circadian disruption, a natural question to have might be: which one matters more? And what about the other numbers I introduced at the start of the chapter, like sleep efficiency? How much do they matter?

DOMAINS OF SLEEP HEALTH

Sleep as something multidimensional makes sense. You could be hyper-regular but not sleeping enough, and that would be bad. You could be hyper-regular and sleeping enough, but wake up every five minutes during the night, exactly on the minute. That would also be bad. Or you could have tons of long, unbroken sleep but have it happen randomly around the clock, which would mean low sleep regularity and a disrupted circadian clock (bad).

So rather than look for *THE* sleep thing to care about, we can instead think about multiple important aspects of sleep at once: how long you sleep, how regular your sleep is, how often you wake up in the middle of the night, and other dimensions related to our understanding of sleep health. I already mentioned "RU-SATED?" (**R**egularity of sleep, **S**atisfaction with sleep, **A**lertness during waking hours, **T**iming of sleep, **E**fficiency of sleep, and **D**uration of sleep)—you could also add things like "how easily you can fall asleep and return to sleep," or "number of naps per day," or "rhythmicity," which is often defined to be "how much your daily activity pattern resembles an up-and-down wave shape."

We've started using multidimensional sleep definitions to ask questions like "How do these, taken all together, predict [bad thing]?" or "Which of these individual components of sleep health best predicts [bad thing]?" Already, the results are compelling: Multidimensional sleep, taken as a whole, predicts mortality better than many other things you'd expect to affect your lifespan—heart attacks,

---

34 *In practice, they're often weakly correlated—e.g., wildly irregular people are slightly more likely to have lower sleep durations—but theoretically they capture two different things.*

35 *Not recommended!!*

depression, a history of stroke, a history of smoking. It also predicts mortality better than your own self-rating of overall health does.

When you dive in a little deeper to see *which* parts of sleep health seem to matter most, sleep duration rarely floats to the top. In one study looking at mortality in older men, rhythmicity of sleep and WASO were the two most important sleep-related factors for survival. Sleep duration was only about one-tenth as important as each of those metrics were.

Another study, in a larger, more diverse sample of individuals, found sleep regularity to be a better predictor of mortality than sleep duration, with worse sleep regularity being reliably linked to worsened outcomes.

Rhythmicity and sleep regularity tend to be correlated, though rhythmicity is more about your activity patterns over the whole day, while sleep regularity is specifically about the timing of your sleep. But both are scores that get at the health of your circadian rhythms and the robustness of your sleep groove by extension. High rhythmicity means you look like you're in a groove. Low regularity means you're not in one.

WHAT'S GOOD SLEEP, THEN?

Thinking of sleep in this multidimensional way is great; it's the future. But one challenge with multidimensional sleep as we work with it today is that it's not snappy to communicate to someone what *is* healthy. You can't just say "seven to nine hours a night" and ride off into the sunset like you can when you're talking about sleep duration. What are the *thresholds* for good health with multidimensional sleep? How can you know if your sleep is unhealthy?

I think we'll probably pare down the dimensions we talk about as a clearer picture of the most important ones emerges in the coming years, and that will make things simpler. But until they do, and even after, I like the framing of being in a sleep groove, which implies an effortless rhythm with momentum behind it.

- Are you doing pretty much the same thing every day? Rhythm.
- Do you feel confident that you'll fall asleep at the start of the night and fall back asleep if you wake up in the middle of the night? Effortless.
- Do you normally feel good about your sleep—and do you feel like one bad night of sleep wouldn't throw you off too much the next night? Momentum.
- Do you feel like you're not in a groove? Then you're probably not—no number needed.

SLEEPING IN ON THE WEEKEND: GOOD OR BAD?

This is a question that gets a lot of attention, and one that ties into our discussion of multidimensional sleep health. After all, sleeping in on the weekend is a trade: more sleep duration, less sleep regularity.

Here's an argument *for* sleeping in on the weekend:

> *There's no evidence that your body sleeps more than it needs to, but there is evidence that not getting enough sleep can impair your performance and damage your health. If not sleeping in on Saturday means that you're averaging six hours of sleep per night instead of seven or eight, then you're engaging in a behavior that's been associated with increased health risks. Plus, in a large-scale study looking at sleep patterns and mortality, people getting five hours of sleep a night during the work week were able to "rescue" themselves from the mortality risks of such a pattern of behavior by sleeping in on the weekend. Isn't rescuing yourself from dying early a good thing?*

Here's an argument for *not* sleeping in on the weekend:

> *Sleeping in late on the weekend doesn't just clear out some accumulated sleep debt; it also delays your body's internal clock by making you miss out on light in the morning. This shifts not just your body's sleep rhythms but also your non-sleep rhythms, like glucose processing. Plus, it can allow sleep disturbance to propagate forward to the next day. When you sleep in on Saturday morning, you're making it more likely that you'll stay up on Saturday night, which can delay your sleep even more on Sunday, until you crash awake Monday morning, jet-lagged and miserable.*

Here's an argument I would make for a third way, trying to make everyone happy:

> *"Sleeping in" vs. "not sleeping in" is a false dichotomy! You can sleep more on the weekend by simply going to bed early. If you're chronically sleep-deprived during the work week and you go to bed early Friday night, you may still sleep in the next morning, but it likely won't be as late as it would be if you hadn't gone to bed at the earlier time. This means you'll recover from lost sleep without taking quite so big a hit to your rhythms.*

If you're rolling your eyes and thinking *I get five hours of sleep a night during the week, but there's still no way I'm going to give up staying up late for fun on Friday,* I hear ya. Sleep in on the weekend all you want; it's probably the better choice for you.

But take this reminder on your way: going to bed at your normal time or later and sleeping in on the weekend does still come with a cost. It pays off sleep debt at the cost of throwing off your groove and making it less likely that your sleep in the following days will follow the regular pattern of the week. It does, in fact, trade sleep regularity for sleep duration.

REGULARITY VS. DURATION

If you were to hold me at knifepoint and demand, "Which is more important?! Sleep regularity or duration?" I would give you the cop-out answer of "Well, it's not either-or." Of course it isn't! You can sleep enough *and* do it at the same time every night. *Can't we all get along?*

If you put the first knife away and took out an even bigger, more threatening knife, I'd say, "Okay, okay, okay." Then I'd say this:

1. If you're truly not getting enough sleep per night, to the point where you or the people around you think it's affecting your ability to function, work to increase that number first. But try to increase it every night; don't settle for boosting it by only sleeping late into the morning on weekends.

And:

2. In multiple studies, the relationship between mortality and sleep regularity has been stronger than the relationship between mortality and sleep duration. There's a clearer picture in these studies that sleep irregularity is bad than there is that moderately short sleep durations are bad.

Sure, the reason for this could be *entirely* that sick people with long sleep durations are masking the sleep duration trend. And not getting enough sleep for your body is *still bad,* regardless of regularity. But we've known about eight hours a night for a long time now, and it's increasingly clear that "eight hours a night" as a definition of good sleep is not enough. Multidimensional definitions of sleep health are the future, but it's a lot to track at once. Regularity is straightforward, and there are good, intuitive reasons to care about it. Let's care about it.

WHY YOU SLEEP WHEN YOU SLEEP, OR THE TWO-PROCESS MODEL

I have a beautiful collection of analogies for the physics of circadian rhythms; they are my treasures:

1. Circadian rhythms are like *being on a swing,* where a push in the outward direction while you're moving forward is like light exposure during the day or morning, and a push in that same direction when you're swinging backward is like light exposure in the evening or dead of night.

2. Circadian rhythms are like *walking,* stretchy and flexible, where you can adjust the size of your gait bigger or smaller so that you can walk two steps per sidewalk square, if that's the kind of thing that brings you joy.

3. Circadian rhythms are like *bouncing with friends on a trampoline*, where colliding signals can either communicate a clear sense of up and down or an ambiguous sense, depending on how synchronized they are.

Sleep is a bit harder to cleanly analogize. We often talk about a "hunger" for sleep that builds up when you're awake and drains when you're asleep. This is also called your "sleep homeostat," "sleep homeostatic drive," or "Process S."

It looks like this:

There are plenty of straightforward analogies for sleep hunger; for instance, *being hungry*. Here, being awake is building up an appetite, and sleeping is letting yourself feast. Equivalently, it's like exhausting a muscle from overwork and needing to rest it. You keep going until you can't, and then you stop for a while and recover.

Yet it's not *just* this hunger that drives our sleep. Circadian rhythms interact with sleep by *promoting sleep* at some times and *promoting wake* at others. This is also called the "circadian drive to sleep" or "Process C." For day-adjusted folks, this circadian drive tends to send its strongest signal for sleep in the middle of the night.

Maybe the most concise way I've come up with for capturing this idea is the following picture:

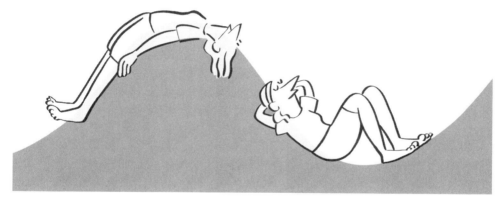

"THERE ARE TIMES WHEN YOUR BODY'S CLOCK MAKES IT HARDER TO FALL ASLEEP AND TIMES WHEN IT MAKES IT EASIER."

Specifically, there are times when your circadian clock wants you to sleep and it's easier to, and there are times when your circadian clock doesn't want you to sleep and it's harder.

The combined effect of your homeostatic and circadian drives to sleep is your net sleep drive. This is called the *two-process model of sleep*. You can think of these two signals as getting added on top of each other to yield your overall urge to sleep.

After all, that's what happened in the extreme sleep deprivation experiments:

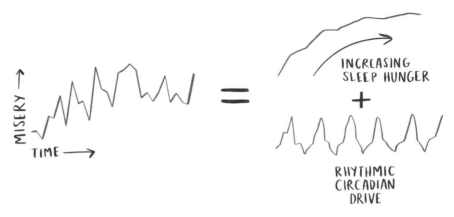

So your circadian clock and your hunger for sleep work together to determine the times you fall asleep and the times you wake up. We often visualize this phenomenon like this:

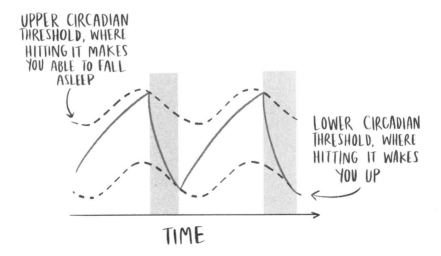

The zigzag is your sleep hunger—the homeostatic drive for sleep. The upper and lower thresholds, wiggling daily, are set by your body's circadian clock. When the sleep drive hits the upper threshold, your body switches into "wants sleep" mode and tries to get you to go to bed. When it drops below the lower threshold, your body switches into "wants wake" mode and tries to get you to wake up. Hit the upper limit, fall asleep. Hit the lower limit, wake up.

You can start to see interesting sleep phenomena simply by changing the alignment of your sleep hunger and your circadian rhythms:

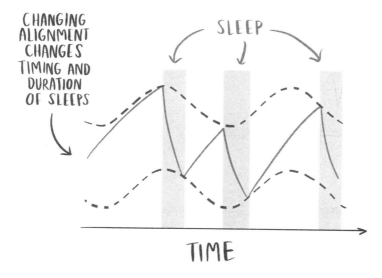

CHANGING ALIGNMENT CHANGES TIMING AND DURATION OF SLEEPS

SLEEP

TIME

These kinds of shortened, round-the-clock sleeps are similar to the kinds experienced by shift workers, recent travelers, and people who aren't in a sleep groove. You don't need a complex model to explain them—the two processes of sleep hunger and circadian rhythms are enough.

So my analogy for the two-process model combines two things: sleep is like a *water cooler* on a *swing*. [36]

The water cooler, here, is capturing sleep hunger. You fill the water cooler during the hours you're awake. You can stay awake even when your water cooler is very full simply by keeping the tap closed, just like how humans can stay awake for long periods of time despite being very, very sleepy. Pressure on the tap from the water inside is like the sleep pressure we feel when we're tired.

Trying to fall asleep is like opening the tap to let water flow out. If the cooler isn't full enough (you don't have enough hunger for sleep), then there's not enough water to reach the level of the tap and you won't be *able* to fall asleep, despite your best efforts. But just like you can tilt a water cooler toward yourself to pour yourself a glass when the water level is low, so too can you fall or stay asleep when your sleep hunger is low, as long as your body's circadian clock is *tilting* in sleep's favor.

---

36   *If you've never put a water cooler on a swing—*

This is where the swing part comes in:

WATER
ABOVE TAP =
CAN FALL
ASLEEP

WATER AT TAP
LEVEL = CAN
FALL ASLEEP

WATER BELOW
TAP = CAN'T
FALL ASLEEP

Over the course of a normal night of sleep, you drain your sleep hunger until you reach the point where you would wake up (drop below the line of the tap) if sleep hunger was the only thing that mattered. But roughly around the time of CBTmin—3:00–4:00 AM-ish *on average*—your circadian clock sends its strongest signal to stay asleep, tilting things so that you can stay asleep (drain a little more). Then, as it starts to swing back the direction it came from, you lose that tilt and wake up.

We can talk about plenty of sleep phenomena you may have experienced through this lens:

- **TRYING TO FALL ASLEEP EARLIER THAN USUAL AND CAN'T?** That's like a water cooler that's full but tilted away from you (your circadian signal is still on the "day" part of its rhythm), so nothing can drain.

- **WAKING UP EARLIER THAN YOU WANT?** Your circadian clock is swinging forward into daytime, tilting away from sleep, causing your sleep hunger to drop below the level needed to drain.

- **WAKING UP IN THE MIDDLE OF THE NIGHT?** That's like a water cooler that's either not full enough (say, you took a late-in-the-day nap) or not tilted toward you enough (your circadian signal is mistimed or dampened).

- **DRANK A BUNCH OF COFFEE LATE IN THE DAY AND NOW YOU CAN'T SLEEP?** I don't know; that's like sticking chewing gum in the spigot so nothing can flow out. Listen, no analogy is perfect.

It's straightforward to see how you could both sleep *against* your rhythms . . .

MORE THAN ENOUGH SLEEP
HUNGER TO FALL ASLEEP
EVEN THOUGH YOUR
RHYTHM IS PROMOTING WAKE

. . . and be unable to sleep, *despite* your rhythms wanting you to:

NOT ENOUGH SLEEP HUNGER
TO FALL ASLEEP EVEN
THOUGH YOUR RHYTHM
IS PROMOTING SLEEP

I can keep going: The middle-of-night wave of sleepiness experienced when you stay up all night? Think of that as the time when your circadian clock is most tilted in favor of sleep, putting maximal pressure on the tap.

The reason we can't clear all our sleep debt in a single day? Your circadian clock is swinging forward into day mode and cutting off the flow, even though there's still plenty to drain.

What about the latter half of the Long Dark experiments, where people had lots of time available to sleep but weren't falling asleep immediately nor sleeping continuously throughout the night? In a sense, the extended sleep opportunity in that experiment meant that those people had the taps on their water cooler open for much longer than most of us do. They were draining their hunger for sleep to such an extent that it would start to sputter out in the middle of the night, leading them to wake up for a while until they refilled the tank enough to have something to drain again, or the cooler tilted back in favor of sleep.

In this analogy, someone with a shallow swing—a low-amplitude circadian rhythm—will be more likely to be able to fall asleep around the clock (no times when the cooler is tilted strongly away from them), but they probably won't be able to sleep for a long period at a time (no times when the cooler is tilted strongly toward them). These people won't be in a sleep groove.

Similarly, someone whose rhythm is more robust but not well-aligned with the rising and draining of their hunger for sleep may find their ability to fall and stay asleep frustratingly hard to predict. This, too, means they're not in a groove.

Circadian disruption, like the kind introduced by unexpected light and activity at night, can make it so your clock is tilted against sleep when you want it to be tilted toward sleep, or simply not tilting enough at any point of the day. Yet one of the great things about circadian rhythms is that they can change. The fact that our clocks can be disrupted also means they can be fixed. More on this ahead.

ALERTNESS LOW, ALERTNESS HIGH

The waves of sleepiness experienced by the participants in the 1968 sleep deprivation study, where they felt better, then worse, then better, come for free in the two-process model: their sleep hunger was increasing the longer they were awake, while their circadian drive to sleep was rising and falling during each successive day of sleeplessness.

That peak in the circadian drive to sleep is the time when your body's clock is most desperately trying to get you to bed. Rather famously, this period of low alertness has been linked to a number of major disasters, including Chernobyl, Three Mile Island, and the *Exxon Valdez* disaster. It's also been implicated in a number of airline crashes, to the point where pilots are warned about the "window of circadian low" on long-haul flights.

If there's a period of peak "being bad at things," is there a period of being good at things as well? I'd say there's two.

First, there's a period in the morning where we've drained our hunger for sleep and our circadian clock is now beginning to stop promoting sleep quite so much. For many of us, it's in the midmorning. It's not likely to be *first* thing in the morning because the circadian clock hasn't fully moved away from promoting sleep and because of *sleep inertia*, the short-term grogginess we feel upon waking up that usually passes in about thirty minutes. But it will be pretty close to first thing in the morning because draining your hunger for sleep is a pretty effective way of getting your mental agility back.

As you go about your day, you'll build up that hunger for sleep again, which will gradually make you worse at things again. This accrual of sleep hunger is offset by a second period near the end of the day called the "wake maintenance zone." I think of it as the "get your act together before it gets dark" part of the day. The wake maintenance zone is marked by a surge in alertness which could have, evolutionarily, existed to give us the energy to do anything important we needed to do before the sun went down. Whenever someone tells me they do their best work at night, I think, *Oh, during the wake maintenance zone.* [37]

We can think of midday dips in alertness as something of a handoff between the morning alertness surge and the wake maintenance zone. A midday wave of sleepiness can arise from the accumulation of sleep hunger that isn't yet offset by the late-in-day wake-maintaining circadian surge in alertness. To use the water-cooler-on-a-swing analogy: Your tank is getting full, putting pressure on the tap. Your rhythm will soon tilt the cooler away from you, relieving that pressure, but it hasn't done it yet.

That's not all there is to midday dips (source: I am a human who has overeaten at Indian buffets for lunch before), but it's an explanation for why midday dips arise both independently of food and somewhat spottily in controlled research studies. After all, interindividual variations in sleep hunger accumulation and circadian timekeeping could make it so that not everyone experiences a dip at the same time, or at all.

The wake maintenance zone is one of the reasons why, if you're trying to fall asleep earlier and become more of a morning person, getting into bed early is a great way of burning yourself out. Imagine you habitually go to bed at 1:00 AM and decide you're going to shift yourself to a 10:00 PM bedtime. You climb under the covers at 10:00 PM, put on your Ebenezer Scrooge–style sleeping cap, turn out the lights, and feel *wide awake because you're in the middle of your wake maintenance zone.* You keep lying there, determined to force yourself into sleep through willpower alone, all the while building up an association between being in bed and stress. You fall asleep even later than you normally do and sleep in the next morning past your normal wake time.

There you go, you think, *I'm doomed to forever be a night owl.*

---

37   *I also think, "This person might have a delayed circadian clock due to their self-selected light exposure, which makes it so they feel increased grogginess during the first part of their day, causing them to feel as though their only period of high alertness is at the end of the day, when in fact they could have two periods of increased alertness if they were more attentive to their light exposure at night."*

Or maybe you just frustrated yourself by waiting for a sleep bus for hours at a stop with a huge "LINE NOT IN SERVICE" sign posted.

So getting into bed during the hours when your circadian clock is most tilted in favor of wake is not a great way of shifting your rhythms to be earlier. Want to know *better* ways of doing it? Read on, my friend.

LIGHT IS A DRUG

ONCE I WAS TALKING TO a guy at the gym about light exposure and its importance to health, and he was nodding along the whole time. He seemed really into it. This made me get a bit puffed up about the whole thing.

"And that's why light is no less than a drug—a drug we ingest through our eyes," I declared.

"Completely agree," he said.

"And we walk around oblivious to how we're dosing ourselves," I went on. "Like the women working in radium watch factories in the 1920s were oblivious to the toxic effects of the radioactive paints they were using."

"Totally," he agreed.

"We'll look back on photos of people awake at night with harsh overhead lights on the way we look back at pictures of newspapers describing heroin as a wonder drug cure-all. We'll realize how having the lights on like that was like having ourselves hooked to an IV drip of poison," I finished with a flourish.

"I also don't think people need to eat food," he said, still nodding.

"Wait," I said.

"If we went to the top of a mountain and soaked up the sun, I think humans could photosynthesize," he continued.

"Hang on," I said.

Light exposure, for as much as its literal chemical effects on the body are incredibly well established, gets a bad rap. Or rather, it barely rises to the level of getting a rap at all. People just sort of tend to feel like it can't matter *that* much. It's not a pill, like melatonin. It's not something you buy at a store. It has no mass. The sun is free.

For some people, it's worse than believing light is ineffective. Light therapy, to them, seems squishy, a pseudoscience. Two steps away from magic crystals and energy vibrations. Probably people like Gym Guy (who, I swear, *genuinely did appear to believe* that humans as a species can all just stop eating and start absorbing the sun for nutrition) don't help with this perception.

But even though I was very thoroughly on my soapbox when I was talking to Gym Guy and maybe had my head in the clouds as I was doing it, I really do believe those things I told him about light exposure. I do think the idea of light poisoning will make sense to future generations and that our cavalier attitude toward light right now will look cartoonish and grim in fifty years.

When light hits our retinas, the photic signal converts to a chemical reaction, which triggers more chemical reactions, which trigger an electrical response to pass the signal along to the brain. The light you're looking at as you're reading this is actively zapping your brain.

And it's not just zapping the parts of your brain that you're consciously aware of, like the ones responsible for processing what you see. It's zapping your *subconscious* visual system—the parts of your brain that adjust your pupil constriction, modulate your mood, suppress the production of melatonin, and, you guessed it, set the time of your circadian clock.

It makes sense for these to be subconscious processes since we didn't used to have much control over the ambient light exposure we got. If the light was largely running on autopilot, (i.e., came only from the sun) our bodies could run on light-processing autopilot as well. Why divert computing power from important visual processing (where to step, where to look for food, what to run from) in order to manually adjust the diameter of your pupil? Why make your brain's "time zone" something under your active, conscious control? If the pattern of the sun is always the same, day in and day out, then why can't systems in charge of processing it be set-it-and-forget-it?

Except the light most of us get nowadays is emphatically *not* the same every day. We can keep the lights on as late as we want at night, and we can block light in our bedrooms until deep into the morning. Shift workers can work 7:00 PM to 7:00 AM on Tuesday and 7:00 AM to 7:00 PM on Friday.

Even if you do get the same light every day, the difference between our days and nights has been worn away: the days we experience are much less bright, thanks to time spent indoors or out-of-sync with the sun, and the nights are much, much brighter than they were when our default, subconscious visual system settings were first locked in.

It's like our subconscious visual system is a traffic intersection where the timing of the lights was set exactly once, right when it was built, and never updated to account for time-of-day, rush hour, and changes in traffic pattern. The sluggish adjustment of our circadian clocks to new time zones is like a backlog of cars trying to force their way through an intersection designed to handle two 1941 Oldsmobiles per hour. We're overloading a machine that was built for simple, repeatable inputs with wildly irregular, round-the-clock stressors.

WHAT YOU SEE WITHOUT KNOWING YOU'RE SEEING IT

For a long time we thought there were two types of "photoreceptors" that could process light in our eyes: *rods* and *cones*.[38] We knew about rods and cones going back to the 1700s—we could see them using lenses from some of the earliest microscopes. Animals without them were blind, which was pretty compelling evidence that they mattered for vision, and for a long time "rods and cones are the only photoreceptors" was the orthodoxy.

There were clues that this wasn't the whole story. In 1927, ophthalmology researcher Clyde Keeler observed that rodless, coneless mice—blind in all understood senses—still appeared to adjust their irises in response to light. But, as can happen in science, with no tidy explanation on hand to explain the phenomenon, Keeler's findings were put to the side as a physiological curiosity.[39]

Flash forward to the late 1990s, when researchers at the University of Virginia uncovered a new type of opsin (that is, a receptor capable of responding to light) in the skin of frogs. Called melanopsin, the opsin was soon after found to exist in the mammalian retina as well. But it wasn't on the rods and cones. It was attached to a

---

38 *Really four, because there are three types of cones.*

39 *Probably, if I had been a scientist at that time, I would have secretly believed he just messed the experiment up and his mice weren't actually as blind as he thought. "They probably weren't fully blind," I would have stage-whispered to my neighbor in the old-timey science auditorium he was presumably presenting his results in.*

special subset of an entirely different class of cells, retinal ganglion cells. The special retinal ganglion cells that had melanopsin—and therefore the ability to respond to light themselves, directly, without rods and cones— were thus given the snappy name of *intrinsically photosensitive retinal ganglion cells,* or ipRGCs.

Finding a new type of retinal cell in the early 2000s led to a flourishing of further research. IpRGCs were found to connect to the suprachiasmatic nucleus, the hub of circadian rhythms in the brain. There were different types of them that had different shapes, firing patterns, and seeming functions. But there was one thing they all had in common, thanks to melanopsin: they all loved 480 nm light.

480 WHAT?

If you look online, you'll see some articles saying blue light is terrible for your sleep and some articles saying blue light doesn't matter. There will also be some articles saying blue-light blocking glasses do nothing and others saying you need to tint everything orange after 7:00 PM. Some amount of back and forth comes with the territory in science, but when it comes to blue light, it's made worse by the fact that talking about light in a precise way is really hard.

When I think about light, I don't think about words like "blue" or "orange." I think about a mountain range. The mountain range is the spectrum of the light.

Light is somehow both a particle and wave, which is not even what I'm talking about when I talk about light being hard to talk about.[40] The important thing to know is that there are lots of particles-that-are-also waves happening when you exist in a room with lighting, and each one of those waves is a rhythm with its own period. Instead of calling it period, though, we call it "wavelength," lest keeping track of the terminology should become too easy for us.

If you put me in a room and gave me a very expensive photodetecting device, I could tell you what the spectrum of that light was. The spectrum is "how much of each wavelength there is." If there's a lot of one particular wavelength in the light, there's a spike in the mountain range. If there's roughly the same amount of each type of wavelength, the mountain range is pretty flat.

---

40 *Don't think about it! That just makes it harder!*

Sunlight, as a mountain range, looks something like this:

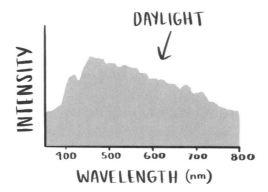

Other types of lighting look like this:

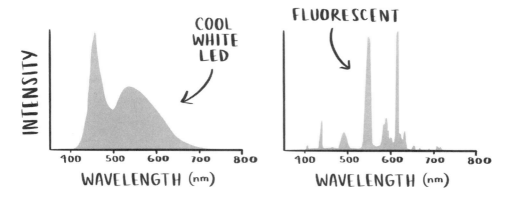

There's a ton of information in a spectrum! If I wanted to write down what the spectrum of sunlight is, I couldn't do it with just a handful of numbers. Even restricting myself to the range of wavelengths that are visible to the eye (i.e., ignoring infrared and UV light) still leaves a range extending from a wavelength of roughly 400 nm to a wavelength of 700 nm, with a corresponding "intensity" level at each.

Writing down how much 400 nm light there is in sunlight, and how much 401 nm light there is, and how much 402 nm light there is, and so on, all the way up to 700 nm would take me three hundred numbers, and that's just an arbitrary way of slicing it! If I wanted to be even more accurate, I'd have to break it down into even finer divisions (e.g., 400.0, 400.1, 400.2, . . . 699.9, 700.0), which would take me even more numbers to express.

Our retina doesn't have input slots to process three hundred numbers, let alone the full information in a spectrum. We have rods and cones. There is one type of rod and three types of cones. Light is this complex, multidimensional mountain range, and our photoreceptors reduce it down to four numbers.

This simple fact—that the best our eyes can give us is a massive, massive reduction of the information contained in the light we're exposed to—explains a lot of the confusion around blue light. There's not *one true* blue light.

Pretend for a second that I told you to capture the shape of the mountain range with four horizontal lines as some sort of zen exercise in minimalism.

E.g., I gave you this:

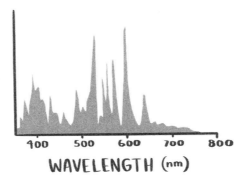

And you had to do your best take on it; something like this:

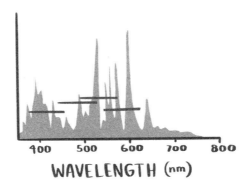

This is functionally what your cones do when they "see" a color. They kind of average the heights of the mountains in a chunk of the range to give an idea of what's going on there.

The net color you consciously experience at the end is set by the ratios of what each cone type sees to one another, like mixing paints together in different amounts. The color you get at the end isn't set by how much red paint you pour in, but how much red you pour in *with respect to* how much yellow and blue paint you poured in at the same time.

The thing is, you're throwing away information when you do this kind of averaging step. You could start with two very different spectra but end up with the same four minimalist lines describing both of them. This means you can have very many, very different spectra that all look like the same "color" to your conscious visual system.

And now we get to the punchline: ipRGCs—your subconscious visual system—love 480 nm light. That's a tiny slice of the mountain range. If you made an LED light source that was just a little spike around 480 nm by itself, it would look blue-green to your eyes.

So you could have a spectrum that looks "blue" to your eye with a *ton* of 480 nm light, which means it would make your ipRGCs go wild. They'd fall over themselves for a blue containing lots of 480 nm. Yet a different blue, *which could look identical to your eyes,* might have essentially no 480 nm light, which means it wouldn't be expected to trigger much of a response from your ipRGCs at all. [41]

This is called metamerism, and lighting that looks "blue" but doesn't actually have a lot of representation in the traditionally "blueish" parts of the light spectrum can be called metameric lighting. Metameric lighting is what could ultimately make it so our screens, lamps, and TVs look blue to our conscious visual system (rods, cones) and mostly dark to our subconscious visual system (ipRGCs).

Most lighting these days, though, is not trying to tell one story to your cones and a different story to your ipRGCs. If it looks bright white or blue, it probably has 480 nm light floating around in there somewhere. That said, if you see a headline that says, "Blue Light DEBUNKED," it almost certainly means "the experience of consciously perceiving a color as 'blue' is immaterial to the signal received by your circadian clock," and not that 480 nm light doesn't matter.

Since most of us don't have fancy photodetecting devices on hand to check the spectra of the light we're exposed to, the safest things you can do to avoid stimulating your ipRGCs at night are dimming the lights to a very low brightness, avoiding overhead illumination, and (for the hardcore) wearing glasses that explicitly block out the parts of the lighting spectrum that ipRGCS love.

---

41 *Because nothing is ever easy: rods and cones are themselves connected to ipRGCs, which means ipRGCs will be affected by rods and cones independently from melanopsin. The field is still ironing out the conditions under and extent to which this matters.*

A QUICK NOTE ON BLUE-BLOCKING GLASSES

The blue-blocking glasses we use for participants in our sleep and circadian studies don't look cute. They cost about ten dollars and are huge and orange. No one, to my knowledge, has ever asked us if they could take them home with them.

There are also lenses for sale that claim to block blue light all the time. You can get these as add-ons from manufacturers. There are two problems with these clear-looking types of blue-blockers. First, many of them don't actually block all that much light, especially when you wear them *out into the sun, which is incredibly bright*. Second, as you've read many times by now, you do actually want circadian-stimulating light during the day.

Remember, having a big day/night differential is a huge part of getting into a sleep groove. If you're in the dark all the time, it's not too dissimilar from being in light all the time. You don't have a pattern to lock onto, a conductor to follow. You start to drift. Blocking the 480 nm light that feels "loudest" to your circadian clock during the day can be actively *bad* for your rhythms.

THE SUN IS, IT TURNS OUT, REALLY VERY BRIGHT

Let's double-click on that point I made a paragraph or so ago, which is that the sun, in fact, is incredibly bright. We often use a unit called *lux* to describe how bright something is, [42] and if I'm inside my fairly bright house during the day, it might be, what, 500 lux? Maybe 700 lux during the day, 150 as things wind down? Outside, it could be *10,000 lux*.

We don't usually notice that the light is changing by a factor of ten or one hundred, partially because our pupils constrict, and partially because (as a general rule) our senses are set to process things "logarithmically." By that I mean we're set up to be able to perceive details when things are subtle and the big gist of things when things are obvious.

---

42 *Note that something can be very bright from a lux perspective but still not have a lot of representation in the 480 nm region, the area that melanopsin cares about. When we talk about lighting from a circadian perspective, we often restrict ourselves to talking about brightness in just the parts of the spectra that are relevant for melanopsin and ipRGCs. One unit that gets at this idea is <u>melanopic lux</u>, or mlux. It's an important idea, but also, the sun's spectrum has tons of near-480 nm light, so you can be pretty confident that, no matter what, when you're getting sunlight, you're getting a healthy dose of melanopic lux.*

Hearing is logarithmic. The sound pressure waves hitting your ears at a concert might be many, many, many times stronger than the sound pressure waves hitting your ears when you listen to a quiet meditation track, and you'll certainly be able to tell that it's louder. But you're not like, "Wow, the sound pressure waves at this concert are *ten thousand* times stronger than those waves from last night's meditation." Your brain adapts to the 10,000× multiplier, and in the end, you just hear.

Vision is similar, in the sense that we can see at very low lux and very bright lux, and we know it's brighter, but we don't think about it too much. Which is why I want to do another special callout for the sun.

I'm a huge fan of lighting technology that helps you get lighting that's "good," circadian-wise (after all, I see it the same way you might see a pill regimen, or a healthy diet). But I love, love, love that the sun is free. Making your life better from a circadian perspective doesn't have to break the bank. It can just be going outside.

YOUR IPRGCS GET TIRED

Something to know about the cells in your eyes with melanopsin is that they can send signals for a long time—a really long time. Other retinal ganglion cells act by registering when a thing in your vision changes, alerting you of that change, and then shutting themselves up again until the next change. But ipRGCs can fire continuously for at least ten hours in a lab and presumably more outside it.

This doesn't mean they send the same signal over the course of the full ten hours, though. The light response to your circadian clock tends to look like this:

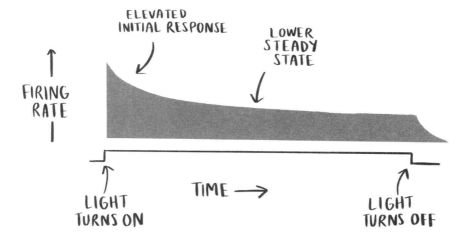

In other words, your body is most prepared to respond to new light after it's been in the dark for a while. Then, after the initial burst of response, it levels out to steady-state. This is similar to how I have run long-distance races in the past: an initial surge of "Yeah, let's do this!" speed, followed by an "Oh God, I have another two hours to go" slowdown after the first half mile. [43]

One consequence of this is that if you're in dim light all day, your eyes wouldn't be quite so tuckered out as they would be if you'd been in bright light all day. These properties of the ipRGCs could be part of the reason why it's been observed that if you get a lot of bright light during the day, you're less affected by light at night.

Another reason could be related to this more theoretical notion of circadian amplitude I keep bringing up: if light during the day boosts the strength of your rhythms, that higher amplitude could help you steamroll right over the disruptive effects of light when your body doesn't expect it.

Or, as in the swing analogy for circadian rhythms:

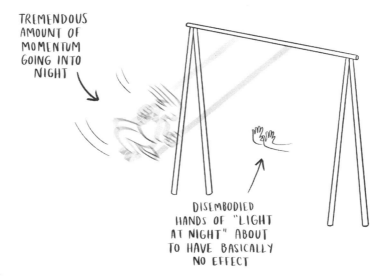

TREMENDOUS AMOUNT OF MOMENTUM GOING INTO NIGHT

DISEMBODIED HANDS OF "LIGHT AT NIGHT" ABOUT TO HAVE BASICALLY NO EFFECT

So in case you needed more of a reason to get light during the day, here it is: light during the day protects your circadian rhythms from light at night, either by tiring out your ipRGCs, giving your brain enormous confidence that day has already happened, or some other third thing we haven't discovered yet. The fact remains that it's generally a good thing to do.

---

43 *See: marathon I was grossly unprepared for in 2016.*

DO I NEED A BIG BURST OF LIGHT
FIRST THING WHEN I WAKE UP?

It's probably not a bad thing, though I wouldn't say it's required. But it *can* be a bad thing, depending on the current pattern of your sleep and how you feel about it.

You can wake up at any circadian phase, especially if you're jet-lagged, a shift worker, or otherwise having severe sleep problems. But if you're relatively well-entrained to a day schedule, you're likely going to wake up during the phase advance region of your circadian day. A big dose of light delivered to your retina at this time is going to fast-forward your clock.

Great. This means you'll probably fall asleep earlier than you would if you hadn't had that light. For most of us—tempted by electronics and media that actively try to seduce us into staying up later—that nudge in the earlier direction is actively good. And if getting that bright light in the morning is part of your daily routine, *continuing* to get light at that time is part of maintaining your sleep pattern. If you *miss* light in the morning, your clock will be running behind schedule by the time bed comes around, unless you compensate for the missed light in the morning by cutting back on light in the evening.

But what if you're someone who chronically wakes up earlier than you'd like? You've tried as much as you can to sleep in, but you just can't. Maybe this has happened more and more as you've gotten older, or maybe it's emerged after a career or lifestyle change, but whatever the reason, you just wish you could convince your body to let you sleep in a little more.

If this is you, light first thing upon waking up is probably the exact wrong thing. Your circadian rhythms are too early for your preference. You don't want to speed them up, you want to slow them down. As counterintuitive as it might seem, the prescription for you is probably to keep the lights on a little longer at night, not blast yourself with light first thing in the morning. This morning dose of light could be making your particular flavor of sleep problem worse, not better.

And it's not just the five minutes or first hour of light you get after waking up that matters. If I gave you a big burst of light first thing in the morning and then kept you in darkness for the rest of the day, your circadian rhythm would not be good—or, at least, wouldn't be entrained to a 24-hour day.

The signal you'd be sending your circadian clock with that bright light pulse once a day is that it's too slow and needs to be sped up. So it would speed itself up. Without any other light information at any other point in the day telling it, "Okay, that's enough, time to cool your jets and slow down," it would go through this speed-up process every single day. You'd likely end up having short day after short day—23.8 or 23.9 hours, instead of reliably 24—and experiencing the same thing blind individuals and others with non24 experience: slowly but surely getting out of sync with the sun.

Of course, you're probably not getting a big burst of light first thing upon waking and then retiring to the blackened abyss to spend the remaining 23 hours of your day. But all of these things exist on a continuum. If you wake up, get a big dose of light, and then sit in a very dim office for the rest of the day, it's not nearly as bad as the single burst of light followed by pure darkness. But it ain't great.

ON DARKNESS

Speaking of pure darkness, when's the last time you've seen any? For me, it was a trip out to a friend's farm last fall. We enjoyed a rustic stroll through their walnut trees, ate their delicious, rustic lentil soup, and moseyed outside once the sun went down to look at some stars, rustically. Then, after about a minute of squinting at the skies, something hit me: my eyes had adjusted as much as they were going to adjust to the darkness, and it was still pitch black. They were wide open, and I could hardly see anything.

I don't normally live in a particularly bright place, but I still don't see that kind of darkness often. There are streetlamps, personal electronics, the light pollution of nearby cities. Even with blackout curtains, I'm still rarely in the dark-dark, as I suspect are a lot of people.

This wouldn't be a problem if our eyes didn't care about the light that was dim below a certain point. If we all just rounded down, a low baseline hum of constant light exposure wouldn't register with our sleep and circadian system. For some of us, this is what happens.

But not everybody. Some people are very, very sensitive to light exposure.

THE EXTREMELY LIGHT-SENSITIVE

How do you know if you're light-sensitive? From a sleep and circadian perspective, it's pretty tedious to find out. You can look at how much your pupil constricts when you shine light on it, but a real functional test is to see how much your body's natural melatonin production gets squashed when it tries to produce it in dim light conditions.

This is what researchers at Monash University did in 2019: They took people on a consistent schedule and had them show up at a sleep lab repeatedly over the course of about a month and a half. On each visit, the researchers had the participants spit into tubes for about five hours to measure the amount of melatonin in their spit over time.

During one of the visits, the participant was kept in total darkness. This was used to calculate their baseline melatonin rise, and for most people it looked like the takeoff of a plane, a ramping-up into night:

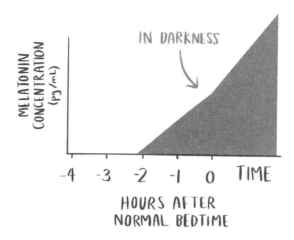

At other visits, though, they kept participants in differing levels of dim light. The light suppressed the subjects' melatonin, meaning that the ramp-up was squished:

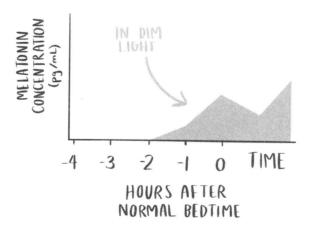

How much it took to squish the ramp-up varied by person. For at least one outlier, it took 400 lux to suppress melatonin by half. That's pretty bright (think: a well-lit indoor space). This person was going to produce melatonin so long as the sun wasn't shining straight in their eyes.

But for most people it took far less light. Some people only needed 10 lux to have their melatonin suppressed by 50%. That's not very bright at all (think: your phone).

There ended up being a fifty-fold spread in the lux it took for people to have their melatonin suppressed by half. Said another way, two people could live in the same house, with the same lighting in their bedroom, and one of them could be fifty times more sensitive to that light than the other.

When the researchers followed up to see how much light people were getting in their homes at night, even the dimmest homes tended to be bright enough that a light-sensitive person would experience significant suppression of their body's natural melatonin in the hours after dark.

Some amount of melatonin suppression will happen naturally if you live at a latitude with late daylight during the summer: the odds are pretty good that if the sun is up and you look outside, your melatonin will be suppressed. But in many other cases, we're doing the melatonin suppression to ourselves by keeping artificial indoor lights on. We're not realizing that we're actively *dosing* ourselves with light exposure.

Light is, in a sense, anti-melatonin. Your body wants to produce melatonin during your biological night, but it won't if the lights are on.

When you think about the billion-dollar melatonin supplement market, it's hard not to wonder how many people who turn to melatonin as a pill could just be turning off their light. Or at least dimming it way, way down. For a whole lot of us, our high light sensitivity means that even a moderately bright indoor light could be mistaken by our biological clock for the sun.

We might not be able to get by with photosynthesis instead of calories, but light is still something we're consuming, for better or worse. And the patterns in which we consume it play a huge role in shaping our sleep and circadian sense of self.

This book so far has been mostly focused on painting sleep and circadian rhythms in broad strokes, setting up the fundamentals. From here on, it'll switch to quicker, lighter strokes, as I jump from topic to topic, revisiting the same basic ideas of rhythmicity and grooves over and over again. If the earlier chapters were like the parts of a textbook where core concepts are introduced, the remaining chapters are like exercises, showcasing how the concepts can be applied to everything from Olympic medalist to night owls. And speaking of night owls . . .

TIME IS FAKE: CHRONOTYPES, JET LAG, AND SHIFT WORK

SOMETIMES I SEE PEOPLE SAY, "I could never get on a schedule where I wake up before 7:00 AM. I'm a terminal night owl."

They're wrong because time isn't real. If I *Truman Show*-ed them—that is, I constructed an entire world for them in a massive TV set where I controlled all aspects of the environment at all times, and I set the sunrise and clocks to be three hours earlier than the actual time zone they were in, and I kept all other aspects of their life the same—they'd eventually adjust to a schedule three hours earlier than the schedule they were on before.

Then I'd pull down the massive curtain around the set and kick in their door right as they were just settling in for breakfast.

"HA," I'd laugh, triumphant. "Who's the night owl now? The real time is 6:30 AM."

You might say this is an unfair example, as I have neither the time, nor money, nor approval from an ethics board to *Truman Show* somebody just to make a point about their sleep schedule. And I should further say that changing the clocks on my movie set to show the wrong time is a bit of a trick too: for someone who's a night owl because they vowed, years ago, to never go to bed before the clock says 3:00 AM, yeah, okay, sure, that person really *is* always going to be night owl.

But most people who identify as night owls don't do so because they've sworn a blood oath to stick to a late-night schedule, come what may. They're night owls because they don't feel tired when everyone else does at night. They might even *like* to fall asleep earlier than they do, if only so they feel less miserable getting up the next morning.

But a person who's an extreme night owl in Portugal (say, going to bed at 5:00 in the morning and waking up at 1:00 in the afternoon) is a pretty *average* person by East Coast United States standards. That Portuguese night owl is getting up and starting their day at nearly the same moment in time as I am in Washington, DC, even though the times displayed on our clocks are different by five hours or so.

A less extreme night owl on the East Coast (falling asleep at 2:00 AM, waking up at 9:30 AM) would similarly look extremely average, if not *early*, if you teleported them suddenly to California, on the West Coast of the United States, where their habits would have them waking up at 6:30 AM instead.

So what's different between these different locations, apart from the times displayed on our phones? Social cues, for one—I'm not getting invited to the club at 4:00 PM eastern time, while the Portuguese night owl might already be out on the town at the exact same moment. But the social pressures of where we live aren't typically what people think of when they talk about being a night owl. Usually, they're trying to capture something more intrinsic to their identity, like a personality type or genetics.

A more important difference, in that sense, is that Portugal, California, and Washington, DC, all have different sunrise and sunset times. Sure, sunrise might happen at nearly the same *local* times in DC and California, but that doesn't change the fact that about three hours will pass between the sun rising in DC and the sun rising in California. People in DC are going to start getting photons coming through their bedroom windows, and then about 180 minutes later the same thing will start happening to people in LA.

What's *intrinsic*, then, is how our bodies respond to these different patterns of light exposure. [44] Give an early bird (or "morning lark") and a night owl the same pattern of light and dark and you'll expect to see the early bird wake up earlier than the night owl. They're processing the same signals in different ways. The time on the clock on your wall has nothing to do with it.

TIME IS FAKE

Which brings me back to my *Truman Show* example. Sure, you can't change the sunrise. But most of us could, if we really wanted to, *Truman Show* ourselves. *Electricity and lamps exist, after all!*

---

44 *along with other timing signals*

I could set an alarm and turn on all the lights in my house at 3:00 AM, faking the sun rising ahead of schedule. I could even invest in special lights engineered to mimic the brightness of the sun and blast those in my retina as I paraded around my house slurping my 3:15 AM cold brew. Then, in the evening, I could become a full-on vampire at 6:00 PM: lights off, curtains drawn, tucked away in the basement, as far from any signs of life and activity as I could get.

If I kept that up for a week or so, I would adjust to an approximately 3:00 AM wake-up time. It wouldn't be magic. I would have just tricked myself into thinking I lived about three hours east of where I currently do right now. I'd be an extreme early bird by the standards of people who live near Washington, DC, but I'd be right in the middle of the bell curve for people who live in the South Sandwich Islands. [45]

So anyone who says they could never adjust to a schedule where they're waking up before X-o'clock *merely lacks the will* to dramatically alter their home, social calendar, and light environment until their body naturally wakes up before X-o'clock on its own. Case closed.

THE TIME ON THE CLOCK ON YOUR WALL DOES ACTUALLY HAVE SOMETHING TO DO WITH IT

Except we live in Society, where lots of things that affect your light exposure are coordinated through collective engagement with the notion of local time. Schools start at 7:00 AM. Workers work nine-to-five. Restaurants open and close at times according to convention, laws, and local demand. The sun, for as much as you *can* avoid it in daily life, [46] will probably still get some of its (very bright) rays into your retina over the course of the day, and solar patterns are usually roughly linked with a local time of noon. There are lots of reasons why the time displayed on your clock can and will affect your circadian rhythms and, consequently, your sleep.

Not to mention we often have *families* and *friends* and people who want to do things with us, people we love and cherish and don't want to chase away at 6:45 PM because it's forty-five minutes past the start of Vampire Time. My husband would not be a fan of my 3:15 AM cold brew parade if it meant noise and shuffling around and light from my 10,000-lux sunlamps risking waking him up at the same unholy time I was trying to adapt to.

---

45   *Found by Googling "place with time zone three hours earlier than DC," but ended up truly enjoying the South Sandwich Islands Wikipedia page and can only recommend it.*

46   *with blackout curtains, working from home, basement offices . . .*

The extremely mechanical instructions for how to stop being a night owl are neither particularly practical nor actionable. They exist as a sort of "sure, you *could* do that, I guess" piece of knowledge, like "sure, you *could* jog to work and spend the rest of the day in your sweat-covered exercise clothes." It's hardcore, and it might help you get your body where you want it to be, but it's also probably not the most sustainable way forward, nor would it be great for your relationships.

But the fact remains: it's possible to do it! It's possible—work hours, family obligations, practicality notwithstanding—to align yourself to whatever schedule you choose! In modern life, we can control our environments, and this means we can move our clock to any time zone we want, without ever needing to get on a plane.

What about *not* modern life? What if we were in an environment without artificial lighting—where the temptation to keep things lit later and later into the night didn't exist?

A STUDY I REALLY WISH I HAD BEEN A PART OF

Here are actual things I've done for studies in which I have either been a lead investigator or a participant:

- Lived for four days in an office with a pullout bed-couch eating only Clif Bars and drinking only Gatorade
- Sat in front of a 10,000-lux sunlamp during the middle of the night
- Gargled cotton swabs in a dark room every half hour, for twenty-four consecutive hours, and slowly developed a taste for them

Here's what other people have done in their studies:

- Gone camping

And proving you don't have to suffer as I have to do good science,[47] the camping studies have had a tremendous impact on our understanding of how our bodies have been affected by the modern replacement of natural light-and-dark patterns with artificial light.

If you sample people's circadian clocks under normal, real-world conditions—say, you bring them into a lab and have them spit into tubes so you can get their

---

[47] *The participants in these suffered in __different__ ways: camping both during a polar vortex and during peak mosquito season.*

dim light melatonin onset (DLMO) on that day—you might get a distribution that looks something like this:

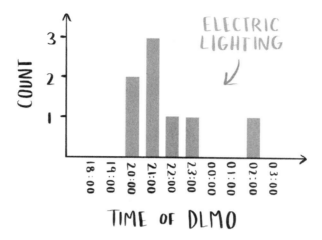

That is, you get a distribution with a long tail. There are more late types—night owls with late DLMOs—than you'd expect, based on symmetry.

But if you take these people camping into the beautiful wilderness of Colorado and forbid them from having phones or screens (campfire and headlamps still okay after dark), and then a week later you return them to society and *immediately* drag them back into the lab to spit into tubes once again, the distribution can change to look like this:

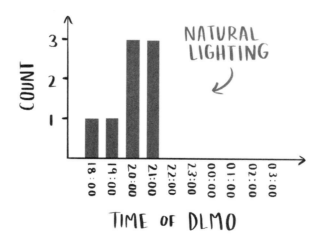

The long tail—the late types who are very, very late . . . vanish. It's still the same people with the same genetics, still living in the time zone they started in. Nothing deep or fundamental has been changed about them, but they became way more of an early type—or at least average person—than they were before the camping. So what *did* change?

1. They got much, much more light, both earlier in the day and throughout.

2. They got much, much less light at night.

Here's a chart from that paper showing how the light they got in their normal life compared with the light they got while camping:

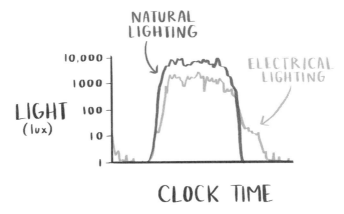

FIGURE 2 from Wright et al., 2013

But this chart doesn't quite underscore how big of a difference there is because it's using a logarithmic scale. Here's what it looks like when you change the axis so that it's no longer logarithmic:

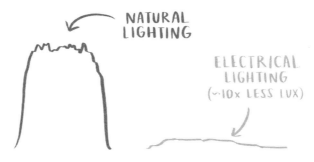

Ten times! Ten times more light during the day when camping vs. not, and a tiny fraction of the light at night when camping vs. not.

In the language of the "being on a swing" analogy for sleep and circadian rhythms, it was like the participants, while they were camping, were being pushed at a nice, regular cadence by someone like this:

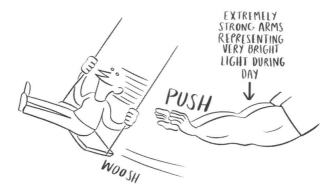

While in their home life, they were being pushed fairly weakly by someone who didn't always move out of the way on their swing backward:

The intuition from this would be that they had much more "sleep momentum" as they swung into their evening while camping. They had more light during the day, which made them more insensitive to light at night, and on top of that, they *barely* had any light at night. Their rhythms moved earlier as a result, tilting their system toward sleep at an earlier hour.

But there's another reason, and it ties back into our discussion of light sensitivity from the previous chapter: you don't notice a difference between a person who is very *sensitive* to light at night and someone who is very *insensitive* to light at night until you put them into some light at night. The at-home, artificial lighting environments had lots of light at night, while the camping environment didn't.

The camping environment was likely a better reflection of the lighting environment we evolved in, and it was highly predictable thanks to the sun being very boring. The sun came up, light got brighter, then it peaked, got dimmer, and vanished. It was like someone playing a piano, going up one key at a time, reaching a highest point, and then going down again. A nice, predictable scale, with darkness as a pleasant, resonant silence separating each day:

 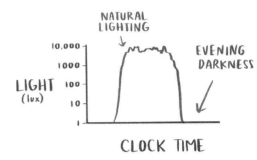

The artificial light environment was *kind of* like that, except that on the way down, right before they hit the bottom note, the key got stuck. Instead of a nice, clean wrapping-up of the musical thread, the last note just . . . didn't resolve. The pleasant, resonant silence never arrived:

 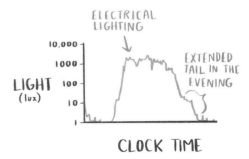

In this analogy, we can think of light-insensitive people as people who could tune out the unfinished musical phrase and light-sensitive people as those for whom hearing the unfinished scale would drive them absolutely bonkers:

Humans evolved in an environment where the light went out when the sun went down, the scale complete. The long periods of light we experience in the evening now, which are bright—not *sun* bright, but not dark—are new.

Differences in how we *respond* to that not-sun-bright-but-not-dark light as individuals probably wouldn't have been too noticeable when we were evolving. After all, if I was highly sensitive to evening light, but was only *in* dim evening light for fifteen minutes of the day, it could only mess me up so much. Moreover, if I was getting super bright light during the day, the difference between day and night would be obvious, and telling time would be easy. But now we live in an environment where we can have dim light as much as we want, as long as we want. And people can get very, very messed up by it.

The camping studies reinforce the idea that some of the most extreme sleep behaviors we see in the real world might not be "intrinsic" to an individual the way we think of them as being. It's not necessarily that you're eternally, irreversibly a night owl. It's that you have the capacity to be a night owl if you're put in a lighting environment that promotes the delaying of your circadian clock.

Or, as I explained it once to my increasingly bored-looking husband: You have two wolves inside of you. One loves getting up early and subsists on early morning light. One loves staying up late and subsists on evening light. Which one determines when you fall asleep and wake up? *The one you feed.* [48]

---

[48] *And if you're a light-sensitive person, it's like your late wolf is a super unpicky eater. It'll eat just about anything you give it, even if it's just a handful of photons. Congrats on your wolf with low standards for what it will eat.*

NO, WE SHOULD NOT DO PERMANENT DAYLIGHT SAVING TIME

If you were reading the last section and you were like, "Hang on. If I'm more sensitive to light *in general,* why does *evening* light matter so much? Why isn't that counteracted by being more sensitive to morning light as well?"—then good for you! This is a subtle but important point.

The short answer is that it's because the effects of evening light on your body can spin out of control in a way that morning light's effects can't, really. This is one of the reasons why adopting schedules that get us more evening light is worse than going on schedules that get us more light in the morning. It's also why we in the United States should not do permanent Daylight Saving Time (DST).

Backing up a bit. A few years ago, I bought blackout curtains for my bedroom. These were pretty long overdue: about thirty feet from my bedroom window is a cheerfully bright, energy-efficient LED streetlamp, which—while great when I'm taking the dog out at night—is the photic equivalent of somebody standing in my azaleas and playing "Seventy-Six Trombones" while I'm trying to sleep.

Getting the blackout curtains definitely improved my sleep by blocking the streetlamp light at bedtime. But I also noticed that they made it so I needed to be even more careful about my *other* sources of light at night. That's because they didn't just block light at night. They also blocked light in the *morning.*

Right now in the U.S., we change times twice a year, with DST in effect during the summer months and standard time in the winter. Basically nobody likes doing this, which is why there's been repeated chatter about making one of them permanent. The more popular option, at least based on legislation that has been introduced in the last few years, is to make DST the time we default to as permanent. This, as a reminder, is the one where the clocks move forward (so it's lighter at night), while Standard Time is the one where the clock moves back (darker at night).

Permanent Daylight Saving Time means not having to change the clocks and not having to experience that gnarly "lose an hour" in the spring. It means no confusion about how many hours offset we are from the time in the U.K. and no struggling to remember if you should say EDT or EST when you're trying to coordinate a Zoom meeting across time zones. As a programmer, I'm generally in favor of anything that makes the totally miserable experience of interacting with dates and times in code even marginally easier.

But it also means—and I'm talking about permanent Daylight Saving Time here—lots and lots of dark in the mornings.

If light did the same thing to our bodies at all times of the day, this wouldn't really be an issue. That's not what happens, though. Recall, from the phase response curve, that light in the morning tends to speed up our circadian rhythms. It tells our internal clock that night is over and it's time to get a move on. It accelerates us toward sleep at night.

And if you get a lot of light in the morning, it eventually advances you forward and forward, up to the point where . . . it stops advancing you. You enter the *delay* region of your internal day, which starts in the mid-afternoon for most people and continues into the early morning. Light during these times is telling your body's clock to take it easy and run slower.

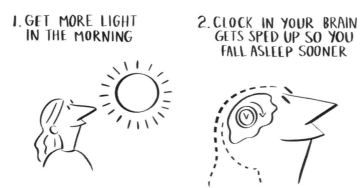

1. GET MORE LIGHT IN THE MORNING

2. CLOCK IN YOUR BRAIN GETS SPED UP SO YOU FALL ASLEEP SOONER

And that slowdown period is the problem. Because while light in the morning is hitting you in the phase advance region, which you eventually get advanced *out* of, light at night is hitting you in the delay region, which is like a temporal sand trap. When you get light exposure in the delay region, your clock gets slowed down, which means you spend *more* time in the delay region. Which means you don't feel tired as quickly, which means you get more light, which means you spend even more time in the delay region. It's a feedback loop that spins out of control.

So if we adopt permanent DST, we're adopting a schedule where we get more light during the hours most people call night and far less light in the hours we consider morning. We're setting ourselves up to fall into the delay region sand trap: more light in the night, making us stay up later and get delayed, and far less light in the AM hours to counteract it.

1. GET MORE LIGHT IN THE EVENING

2. CLOCK IN YOUR BRAIN GETS SLOWED DOWN SO YOU DON'T FEEL TIRED AS EARLY AS YOU OTHERWISE WOULD

3. YOU'RE NOT TIRED SO YOU STAY UP AND GET MORE LIGHT

4. CLOCK IN YOUR BRAIN GETS SLOWED DOWN SO YOU DON'T FEEL TIRED AS EARLY AS YOU OTHERWISE WOULD

and repeat

This is what tanked permanent DST the first time we tried it. I'm not sure why this doesn't always get brought up as the very first point against permanent DST, but we've totally done it before. In 1973, anywhere from 57% to 73% of people supported staying on DST during the winter. So they did it, in January of 1974. By the time February and March rolled around, only 19-30% of people still thought it was a good idea, while 43% said it was actively bad.

What changed? People experienced what happens to your body when you have to kick off your day in the dark of night. They drove to work and caught the bus to school while the sun waited to rise until 8:00 AM. They didn't like it and rolled the decision back before the next winter came around.

You might say, "Well, time is a fake idea. Who says you have to start your day before 8:00 AM?" This is a fair point. We could, societally, shift the normal times we do things to match whatever schedule we wanted. In China, where the entire country is on the same time zone, places like Kashgar (in the far west) have shifted their normal operating hours to reflect the fact that the sun might not come up until 10:00 AM.

But it's a lot tougher to change social standards of when school and work "should" start in every town in the country than it is to pass a bill changing the time that appears on your phone. Which is why we shouldn't do it: permanent DST will put us on a schedule where our traditional social standards for when things should happen are at odds with our biology, sabotaging our sleep and circadian health.

If we want to stop the whiplash of changing the clocks twice a year, why not do permanent Standard Time? I'm in favor of this. It reduces confusion the same way permanent DST does but without the corresponding damage to our internal rhythms. I will fully own that this means some very bright and early mornings in the summer, with the sun coming up in my hometown at (*checks Google*) 4:45 AM. Yet there's a *natural brake* on how much of an effect light in the morning can have on your rhythms, and there's no such brake on evening light.

Permanent Standard Time would also mean that 9:00 PM is dark, even in the summer. But darkness at the right times is a healthy thing. And from a safety perspective, there are lots of streetlamps and other sources of light at night these days that are very good at their jobs.

Which brings it back to me and my window: I need to be more careful about my other sources of light at night because my blackout curtains have meant that I'm no longer woken up by the sun. Sure, I can still wake up naturally in the dark and yank them open myself, like one of the townsfolk in the first song in *Beauty and the Beast*. But if I get too much light at night from non-streetlamp sources, like leaving the lights on overhead in my house, my ability to wake up in the dark in the morning is going to be less reliable, jeopardizing my exposure to the morning light that helps keep me set on the sleep schedule I like. And I'm lucky that there's even morning light to get: with permanent DST, I could be hopping on my first calls of the day while the sky is still black outside.

Social pressures already make it hard for us to get the darkness we need at night (let's face it, screens are fun) and the light we need in the morning. We shouldn't make it harder for ourselves with a change to a system that's already failed once. Permanent Daylight Saving Time is a no-go. Permanent Standard Time? Nice.

WE NOW MUST DEFINE A CHRONOTYPE

The excessive light at night we get in modern life (but not when we go camping), and that we'd get even *more* of if we adopted permanent DST, seems to be one of the reasons we have people who are late *chronotypes*. Yet something I've avoided doing up until this point is actually saying what a chronotype is. "People with greater light sensitivity might be more likely to be a night owl," I've suggested, hand-waving away what "night owl" even means.

This was by design. Talking through chronotype often involves unraveling what people think chronotype *should* mean first, which means it deserves a whole chunk of attention on its own. Here is that chunk of attention.

When I say "chronotype," I usually mean "the halfway point, in local time, between your bedtime and wake-up time on a day when you're not working."[49] Note that there's no mention of genetics here, no "people who have this mutation are chronotype blah."

As the camping study and other research show us, when your circadian markers happen and when your sleep occurs *aren't just* outputs of your genetics. They're outputs of how your body, as a biological machine, has processed the environmental signals you've given it.

Here are three people who might have the same late chronotype:

- A person with a slow intrinsic period, whose biological clock naturally runs at 24.4 (vs. the population average of 24.2)

- A person who is more light-sensitive than average, causing their sleep and circadian rhythms to be delayed by evening light

- A person who is in the middle of the normal distribution for both intrinsic period and light sensitivity, yet lives on the western edge of a time zone and gets more light at night as a result

For the first two, sure, there's a genetic component to those. But in the third case, *that's all environment, baby*. We could do the same thing for the early birds.

Here are three people who might have the same early chronotype:

- A person with a fast intrinsic period

- A person who is light-insensitive and therefore experiences very little phase delay from light at night

---

49   *This is also called the Munich Chronotype Questionnaire (MCTQ), though the MCTQ can also include a correction term for sleep loss during the week. That's a good idea, but we're not going to worry about it here.*

- A middle-of-the-circadian-bell-curve person who's got a smart lamp that blasts them with 10,000 lux every morning at 6:00 AM

One thing I like about this definition of chronotype is that it *also* accounts for the social pressures that shape and restrict our sleep. We don't have to parse out "Ah, this person is not very light-sensitive, and their intrinsic period is 24.2 hours, which is pretty normal, but they do live in *Spain,* which means they're culturally only eating dinner between 9:00 and 11:00 PM, and probably getting light up to and past that time too." We just bundle all of it together and say they're a late chronotype.

Because there is *so much to bundle.* So many things we feed to our body, so many ways the body can parse those signals and be influenced by them. If we want to talk about light sensitivity, let's talk about light sensitivity. But if we want to capture the idea of "Hey, this person goes to bed at what would be considered a pretty late time in their local time zone," let's make that our definition of chronotype. That's what this mid-sleep definition of chronotype does.

WHAT'S JET LAG?

I've mentioned jet lag more than once without really giving a thorough description of what it is. Here's how I think about it: Your body has its own idea of what time it is. It's grooving along according to that sense of time, which is itself constantly being nudged and updated based on the signals you feed it. Suddenly, you do a very novel thing for the human body, evolutionarily speaking, which is to abruptly change the timing of the signals you're feeding the clock. Light, food, and sound suddenly happen five hours earlier or eight hours later because you've traveled five hours east, or eight hours west.

Ah, thinks your body. *I'm super wrong about what time it is.*

Thus begins the process of getting all your rhythms squished, shifted, and realigned to whatever story the new signals you're feeding yourself is telling. It's (swing analogy) like syncing yourself up with the friend next to you on their own swing so that you're moving as one, or (walking analogy) like gradually, over a couple strides, getting your gait locked in with the sidewalk squares again.

Note that I didn't mention sleep. The sleep loss endemic to travel is a bit incidental here. You could sleep hours upon hours every night of your journey and still experience jet lag—it's a phenomenon, first and foremost, of your underlying circadian rhythms.

The focus on sleep disruption is not surprising given that it's one of the most visible symptoms of a disrupted circadian clock. Yet when you define jet lag as "Did you sleep through the night in your new time zone?" you can experience a situation where you sleep great your first night, thanks to an enormous hunger for sleep. Then the next night, your sleep is trash.

"What gives?" you complain. "I thought I had no jet lag!"

What you had, probably, was a crushing, overwhelming pressure for sleep that first night because you had to get up at the crack of dawn to get to the airport and get through security, and the plane ride was miserable and cold, and your seat neighbor took the armrests, and you maybe got two hours of sleep the whole time. Then you drained that massive buildup of sleep hunger with a nice long sleep, and now your sleep hunger is low and your circadian rhythms are getting their say. And their say is that you're still mostly on your old time zone.

I want to make an important point here, which is that time zones are also a fake idea. There's no requirement that you'll *only* shift your clock in the direction you're traveling during your journey; it's all about the timing signals you're feeding your system.

Say I'm going to the West Coast of the U.S. from the East Coast, but I'm taking the earliest flight I can. If I get up at 4:00 AM Eastern Time, that, for me, means I'm getting light and activity during the phase advance part of my day, which will speed my clock up. Except I'm physically heading to the West Coast, which means my theoretical goal is to slow myself down. So I can land in Los Angeles not just jet-lagged from my East to West Coast travel, a three-hour shift, but *super* jet-lagged thanks to the morning burst of light and activity, which have maybe made it so I now need to adjust *four* hours, not three.

In such a scenario, I would probably be fine because I almost never try to adjust to the West Coast from the East Coast unless I'm staying there for half a week or more. I'll doggedly get into bed at 7:00 PM and feel zero angst when I get up at 3:00 AM the next morning and head down to the hotel lobby. Why should I? I'm literally experiencing zero jet lag.

WHAT IT IS LIKE TO TAKE A TRIP WITH ME ACROSS TIME ZONES

Not that I never try to adjust myself when I travel—just that when I do, I'm a complete despot. I try to hack my circadian rhythms every way I can—sunglasses, daylight, eating times, you name it. Here's how my friend Enzo recounted a trip with me when I asked for his honest recollection:

> During our transit time to Helsinki, Olivia made several attempts to indoctrinate us into her sleep cult. She prattled on excitedly using words and phrases none of us understood, such as "circadian rhythms," "blue-light blockers," and "getting a full night's sleep." "Shut up, nerd," we'd warmly

reply, patting her head and side-eyeing each other as family members do when they realize Grandma's no longer 100% there.

We smiled and nodded as she demonstrated specific times she would wear, and remove, her sunglasses.

We waved and dismissed as she demonstrated specific times she ate and managed her body's activity.

And we watched, in jealous, raging, sweaty, oily-foreheaded awe, as she awoke on our first day in Finland at 6:00 AM to the millisecond, with not a single hint of jet lag on her stupid face.

For the next week, as we struggled with 3:00 AM insomnia and 3:00 PM narcolepsy, she awoke and slept perfectly in step with Finland's time zone as if she'd lived there her whole life. It was at that moment I finally realized everything she had been trying to tell us had been true. We were those ignorant world leaders in the disaster movie, finally deciding to listen to the scientist, but only after seeing the meteor incinerate the troposphere.

On our return journey home, we made it a point to listen to her ramblings, except this time, we took it as gospel. And you can imagine my surprise when, after I followed all of Olivia's instructions, on that first day back home, I awoke at 6:00 AM to the millisecond. We would never ever doubt Grandma again.

SOCIAL JET LAG

You might have noticed that physical translocation through space—a real changing of time zones—doesn't factor into my definition of jet lag. Instead, it's the changing of when timing signals occur. You don't have to buy a pricey plane ticket to mess up the signals you're giving your body's clock; you can just stay up late on Friday night. Enter *social jet lag*.

Simply put, social jet lag is jet lag you experience without getting on a plane, often defined by "how miserable your Monday morning is." Literally, the difference in your behavior patterns on Sunday and Monday can be used to compute a score for this phenomenon.

Social jet lag, since it was correctly identified as a thing we should care about, has been linked to all kinds of bad things: reduced sleep, metabolic disorders, depression, etc. It's especially bad for younger people and has been repeatedly linked to worsened performance in high school and college.

Do I think the work being done to catalog social jet lag's ill effects has helped people realize what it's doing to them and curtailed some harmful sleep and circadian habits? Of course. Do I think people are still going to stay up late on Friday nights, even if they know it's bad for them? Of course. But I think that as people become more aware of the *circadian* costs of their actions, and not just the hit they take to their running tally of sleep duration after a late night, they can start to connect that night to *other* things as well, like their mood, or their blood sugar, or their brain fog, and that can start to subtly change their choices in the future.

The more I've practiced a circadian-aware life, the more physically aware I've felt of disturbances to my circadian equilibrium. This makes me great at parties. "Let's just get these guys a little dimmer," I say, as I darken the living room of my friend's apartment at 8:00 PM until I'm badgered to turn them up again so people can read the rules to Codenames. [50] "I'm just gonna—there we go," I say, as I annoy all the other people at the Halloween party by making the lights not just spooky dark, but maybe-trip-down-the-stairs dark.

What I'm trying to say is this: a healthy social life is important, and I'd never recommend someone say no to something valuable and important to them as a social creature just to maintain a rigid kind of sleep and circadian rhythms purity. Like pretty much all good things in life, though, social activities that mess up your body's rhythms—social jet lag—come with a cost. Recognize that the cost is there and you might decide you want to pay it less often.

Or at least you might find ways to soften the blow a bit. Just like I soften the lighting at dinner parties to the point where people can only barely see their food. [51]

SCHOOL START TIMES, OR WE ARE BEING CRUEL TO THE YOUTH

I have at times in my life been desperately, comically sleep-deprived. The double all-nighters I thought were a good idea. The overwhelming feeling of dread as I set my alarm for 4:50 AM after crawling into bed at 1:00 AM.

Yet in terms of actual agony upon waking, nothing to date has beat how I felt waking up for the bus in tenth grade. Getting me out of bed was like scraping a half-baked pancake off a totally ungreased pan. Getting me out the door was like flinging said pancake toward the bus stop, where I'd land, face down, with a squelch. I didn't feel human. I wasn't aware of my surroundings. An hour later, I would be taking the PSAT.

---

50 *a board game*

51 *An exaggeration for comic effect. They can usually see their food okay.*

As teenagers, we have naturally later chronotypes than our parents. Waking up at 6:30 AM for a teen could feel like waking up at 4:30 AM for the me of now. Not only that, we feel more acutely the effects of sleep loss as teenagers than we do when we're older. The tragic irony is that by the time we're old enough to make policy for school start times, most of us have forgotten just how hard it was.

Okay, so how late should school start, then? *11:00 AM,* whispers thirteen-year-old me into my present self's ear. *Make it 11:00 AM.*

Not only would an 11:00 AM start time almost certainly not work from a logistics/teacher-parent willingness perspective, it also probably wouldn't help the kids sleep more. Without *something* to wake them up in the morning (the sun, an alarm), they run the risk of sleeping in. Then they won't have enough sleep hunger to fall asleep at the time they're shooting for the next night, which will make them sleep in more the next day. At a certain point, this stops meaning *more* sleep and starts only meaning *later* sleep.

So what's the right time for school start times? Honestly, it depends on where you live and if you have Daylight Saving Time (or, *gasp,* permanent Daylight Saving Time) in effect. Those factors are going to decide the availability of sunlight exposure over the course of the day, as it relates to local time, a.k.a. the time on the clock on your wall. Even though we filter, block, and otherwise find ways to avoid the sun in day-to-day life, some amount of information about when the sun's around is going to trickle through to our teens. That information, in turn, will affect their circadian rhythms and the time their bodies want to wake up in the morning.

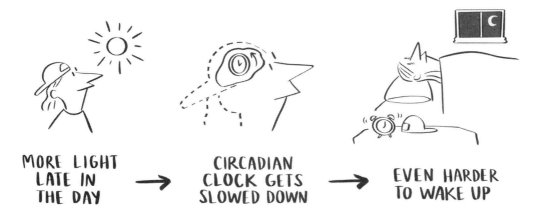

MORE LIGHT LATE IN THE DAY → CIRCADIAN CLOCK GETS SLOWED DOWN → EVEN HARDER TO WAKE UP

As a general rule, though, I'd say, "Do unto teens as you would have done unto your adult self, just shifted by about two hours." That's roughly how much our natural rhythms shift earlier as we go from our teenage years to middle age. If you wouldn't want to wake up at 5:00 AM, don't make a teen have to get up at 7:00 AM, for example—because it will feel like 5:00 AM to them. And then, in like an hour, they'll have to take the PSAT.

IT'S OKAY TO BE A NIGHT OWL (TO A POINT)

There are about ten billion papers showing that night owls have a harder time in life than morning people. They have worse grades, they have worse emotional wellbeing, they're more depressed, they're more likely to die, they eat worse, they gain more weight, they have worse fibromyalgia, and so on.

At least some of these problems are coming from night owls' habits colliding with society and social pressures, as opposed to some inherent failure of their biology. Imagine you've got to take a test at 8:00 AM. A night owl for whom that feels like the crack of dawn will probably do worse than a person who's been up and about for two hours. Or say you've got 6:30 AM training. If your night owl body is still acting like it's the middle of the night, you'll probably perform worse. But the problem in both cases isn't you, the night owl. It's the fact that somebody's making you get up at 6:30 AM.

Yet even if you weren't forced to get on a day person's schedule, you still might have problems as a night owl. A recent paper found that extreme night owls living in sync with their chronotype—that is, not forced to get up early by the tyranny of the early birds—were more likely to develop mental health issues than those who were at least partially dragged onto a day schedule. This could be because extreme night owls are more socially isolated, or it could be because going to bed at 4:00 AM and sleeping until noon necessarily means you'll have less of an opportunity for sunlight during your day. The authors of that paper recommend night owls try to go to bed before 1:00 AM to avoid this increased risk.

If you are the kind of person who prefers a bedtime after 1:00 AM, there's still a bright side: you might be particularly well suited to the demands of night shift work. Which is good, because most people aren't.

SHIFT WORK IS A PUBLIC HEALTH CRISIS

Actual ways shift workers have told me they stay awake on the drive home after work:

- Putting her ponytail into her car's roof when she's driving home in the morning, so that if she falls asleep behind the wheel, it will yoink her head back and wake her up again

- Driving in the middle of two lanes as a rule after her shift, so if she starts to drift off, she won't run off the road

- Calling her daughter, who comes up with things to talk about that keep her mom interested, answering questions, and awake the whole drive back

- Blaring music, blasting the A.C.—anything to make yourself uncomfortable

Not to mention all the people I've spoken to who put their car in park when they get to stoplights because they know they'll let their foot slide off the brakes if they don't. Not to mention that every single one of these examples comes from a person who works in *medicine*.

I once talked to someone in healthcare leadership who said, "I would never send someone home for fatigue. I can't ever imagine doing that. We need their body at their station in whatever form we can get it." Which would be less terrifying if I hadn't also talked to a shift worker earlier that day who yells into nothingness her entire commute home, just to keep herself awake on the road.

Many of us in modern life are on the shift-work spectrum: we don't follow a consistent schedule day in and day out, even though we don't have work forcing us not to. In one study that found sleep regularity was a stronger predictor of mortality than duration, only 10% of the least regular (most likely to die) group was shift workers; the rest were just people with irregular sleep patterns.

So many of us have reasons to care about the health problems of shift workers because those might be our problems as well—problems like depression, obesity, kidney disease, cardiovascular disease, you name it. Shift work is a probable carcinogen. It's terrible for you, health-wise.

It's also a massive safety risk, to the shift workers themselves and to people they care for. Someone too tired to drive an emergency vehicle safely can't effectively save people in need. The people who provide some of our most critical services are risking their own health to look out for the rest of us. Society needs to step up and start treating shift work like the health crisis it is.

HOW TO HELP A SHIFT WORKER

So how do you make things better for shift workers? Where do you start?

One thing you can do is change the shift schedules that people work. Good luck with this. There are truly atrocious work schedules out there, but often these have persisted for as long as they have because there's *some* force keeping them there. Maybe it's the case that the employer needs to distribute the burden of working nights, so the load gets spread across their team, disrupting everyone's circadian rhythms. Maybe people are used to working those hours and like them.

Whatever the reason, I'm skeptical that shift hours will change to be better for workers anytime soon—not because it can't be done or shouldn't be done but because it's hard for organizations to do. You *can* keep shift hours the same but put late chronotypes on later shifts and earlier chronotypes on early shifts—which has shown to be effective in studies where it's been implemented—but at the end of the day, you'll probably still have people working hours they're not particularly suited to.

Which leads us to the second approach: help shift workers manage the health stressors of their work hours by giving them guidance and direction for what to do during their time off.

When you step back and think about shift work in a vacuum, perhaps the best off-time strategy for a nights-only shift worker would be to have them shift their lives entirely to align with their work hours, sleeping during the day even on the days they have off. For a person on this strategy, it would be like living in the United States but pretending you worked the same hours as a person living in Tokyo. With good enough blackout curtains and strong enough willpower to ignore the seductive allure of diurnal life, you truly could fully adapt to a night-living lifestyle. [52]

A tiny fraction of real-world shift workers do this, and those people seem to be pretty well adapted to the demands of their job. But most don't, and the vast majority want to sleep at night during the days they're not working. And the "going full nocturnal" approach would be just as bad as staying on a day schedule for someone who works a mixture of day and night shifts: for those people, jet lag is inextricably baked into their job.

In practice, there's not one "best" strategy for what a shift worker should do during their off time. It's too complicated: There are too many possible shift schedules a person can be on, and too much variation from person to person in how those shifts will affect them. They might also have different goals (sleeping more vs. safety vs. spending time with family, for example).

---

52 *This is just about the only way a shift worker can find a stable sleep groove.*

Still, we can begin to sketch out the space of possible strategies with a few examples.

- If you're adjusted to a day schedule and you've got a one-off night shift tonight before going back to the day schedule, you're not going to be able to meaningfully shift your body's circadian clock in the next eight hours. You're going to want to bank as much sleep as you can in the hours leading up to it and be aware of when your peak fatigue hours are going to occur.

- If you only work night shifts and you don't have a ton of trouble sleeping after your shift wraps up, then you might want to *delay* your body's rhythms: sleeping in late, even on your off-days, with your mornings dim and your evenings bright.

- If you only work night shifts and you find it hard to sleep after your shift ends, then you might want to *advance* your body's rhythms: shifting your rhythms earlier and earlier until you're able to get a solid block of sleep *before* you go to work. This is a strategy that seems to hold promise for older workers, for whom falling asleep at 1:00 PM is easier to adapt to than falling asleep at 8:00 AM.

Even these examples are too simplistic for most people: Advancing or delaying your clock means knowing what your body's personal time zone is—when markers like DLMO and CBTmin are going to occur for you. For workers experiencing the constant, conflicting barrage of signals that come from shift work, this can be hard to pin down. And again: what do you do if you work days and nights on the regular? It's not something a paragraph of text can handle, the same way you couldn't summarize all the knowledge built in to a GPS system in a few sentences.

When words get too hard, math can step in to extend them. Algorithms for helping shift workers are what I work on with my research collaborators (and at my company, Arcascope, too). I'm convinced that algorithms of this kind can help people sleep more, feel better, and thrive more on shift-work schedules—or just typical irregular schedules in typical irregular life.

WHO CARES ABOUT (LOCAL) TIME

Let's bring it all back to time being an artificial construct with one more example. I've sometimes heard people say, "You should go to bed early. Only the sleep you get before midnight counts." As if when the clock strikes 12:00 AM, sleep suddenly becomes shoddy and worthless, a cheap parody of what it was at 11:59.

Of course this isn't the case. What does midnight even mean in an era of artificial lighting, where we don't ever really *have* to see the sun? If you're perfectly adjusted to a schedule where you fall asleep at 1:00 AM every day, and you stay on that schedule all the time, and you get loads of bright light during the day when you're up and about, and you feel fine, and everyone says you're fine too, then you're almost certainly fine. You're not a shell of a human, stumbling through life like a zombie, wishing you'd only gotten some sleep that "counted" the night before.

The issue is that 1) this advice is often given by people who habitually fall asleep before midnight, and 2) when they do stay up close to or past midnight, they're also delaying their body's clock as they do it. Get a bunch of light and activity you normally wouldn't from 10:00 PM to 12:30 AM and you're going to delay your melatonin rhythm. This could make it so you're waking up with more melatonin in your system, which could make you feel more groggy and more miserable.

The real advice is the boring advice: to get into a sleep groove, go to bed at the same time every day. If you can't do that, at least turn the lights down at the same time every day.

BUT IF YOU'VE GOT INSOMNIA

DON'T JUST DIM THE LIGHTS. I've spent this book talking about light, and sleep regularity, and how our bodies are physical engines that convert the signals we give them into a momentum for sleep at night and wake during the day. "Ah," you might say, having read this far. "Got it. Dim the lights at night and I'll be fine."

For some of you, that simple change will be enough! Making your lighting environment healthier can go a long way toward boosting the quality of your circadian rhythms and helping you get into a sleep groove. But if you've got insomnia—capital-I Insomnia, trouble falling asleep, trouble staying asleep, or feeling so bad about your sleep that you've thought about seeking a doctor's advice—just dimming the lights at night is probably going to do basically nothing for you.

It's like putting a Band-Aid on a gnarly gash. Light dimming and other elements of *sleep hygiene* (no phone before bed, etc.) are going to help polish and spiff up your sleep, but if you're truly messed up, trying a simple fix like turning the lights down before bedtime, only to find it doesn't help you, might only make you feel even more burned-out on the whole "sleep science" thing.

Think of it this way: dimming lights in the evening is going to gently massage the shape of your circadian drive to sleep.

If you're mostly entrained to a day schedule in your time zone, the change might look something like this:

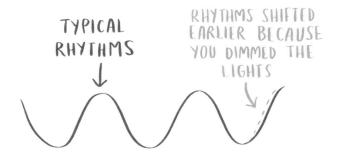

Similarly, the change in your body's natural melatonin production might look something like this:

But if your circadian rhythms are wildly out of sync with your sleep hunger, those changes aren't going to do anything for you. Say you took a late-in-the-day nap. The problem in that case is not that your circadian drive to sleep is 5% weaker than it should be at 10:00 PM; it's that your hunger for sleep is *half* of where you need it to be.

Or imagine you're a shift worker and your body clock is straight-up in the wrong time zone for when you're trying to sleep. If that's the case, then maybe dimming the lights in the evening is the exact wrong thing for you to do. Maybe your circadian clock is running ahead of schedule, and you *want* light at night to help get yourself back on track with the sleep schedule you're shooting for.

Also, and as much as the mathematician in me hates to admit this: *humans aren't just biological input/output machines.* If I was like, "Hey, here's a perfectly dark bedroom to sleep in. Oh, by the way, the bed is haunted by the ghost of a tiny Victorian child who *died* in it," that *might* impair your ability to fall asleep that night, no matter how amazing a sleep groove you're otherwise in. Similarly, if you've had such terrible insomnia for so long that all you think of when you look at your bed is "Ah, the place where I lie awake, miserable, wishing for a sleep that never comes, night after night," you'll probably struggle to fall asleep there.

So light dimming and sleep hygiene aren't bad by themselves; they just might be ill-suited to your particular flavor and magnitude of sleep problem. If you've got a big sleep problem, don't try to fix it with a small patch.

WHAT YOU SHOULD DO INSTEAD

The thing you should do instead is called *cognitive behavioral therapy for insomnia,* or CBT-I. This is a collection of techniques that try to treat the whole of your sleep problems around the 24-hour day and not just the 45 minutes before bed.

The name "cognitive behavioral therapy for insomnia," from a branding perspective, is not great. It's a lot of words, none of which is "sleep," meaning that you might not find it if you search "how can i sleep better please help" at 3:00 AM. It also doesn't really communicate just how effective it is—CBT-I regularly outperforms sleeping pills at managing insomnia.

From the perspective of thinking about things as *having* or *not having* a groove, though, CBT-I can be thought of as cardiopulmonary resuscitation (CPR) for your sleep: Sleep CPR. Just like how CPR works through chest compressions and assisted breathing to help rescue a flatlined heartbeat,[53] so too does CBT-I work to reintroduce rhythms to your sleep. It can help you rediscover a sense of sleep groove.

And like CPR, CBT-I is something you both want to get properly trained on before doing to someone else and don't want to do on yourself. In that spirit, this section is a discussion of CBT-I, but not a How To Guide—go see a professional for that.

Here are the key elements of CBT-I:

1. Sleep hygiene
2. Relaxation techniques
3. Stimulus control
4. Cognitive restructuring
5. Sleep restriction

---

53 *Technically, CPR is more about maintaining rhythms until an automated external defibrillator (AED) arrives on the scene, but "sleep CPR" sounds better than "sleep defibrillation." Please excuse the inaccuracy.*

I've ordered these in what I consider "least surprising" to "most surprising," with the implied importance coming from the notion that if it wasn't surprising, it probably wouldn't be very critical to tell you about.

NOT PARTICULARLY EXCITING OR SURPRISING: SLEEP HYGIENE

Don't look at bright lights too close to bed. Don't drink or eat before bed. Don't look at screens before bed. Be regular with your sleep. Don't drink coffee in the afternoon. Avoid stressing yourself out right before bed.

All of these are functionally placebos for clinical insomniacs for reasons already mentioned. But they're good enough practice for most people, and harmless enough for the others for whom they might not be the right thing to do (e.g., a shift worker or jet-lagged person out of sync with their desired schedule who might *want* light at night to help shift themselves) that we include them in the process anyway.

If we're making the comparison to actual CPR, this might be the equivalent of looking inside a nonresponsive person's mouth to make sure there's not something blocking their airway. It's an easy step to take, and there's no reason not to check it off the list. But just like with regular CPR, you wouldn't expect this step to fix a person on its own if it's not the source of their problems.

NOT SUPER EXCITING OR SURPRISING: RELAXATION TECHNIQUES

If you're blasting dubstep music until 11:59 PM, then getting into bed at midnight in your old-timey Victorian cap and gown, and yet somehow you're finding yourself unable to sleep, allow me to be the first to suggest: don't do that.

Relaxation and mindfulness strategies can be immensely powerful for calming down an overactive brain. Having a soothing ritual you follow in the evening, practicing meditation, or trying biofeedback techniques that help you really *get* how slowing your breathing changes your heart rate can—and do—help people fall asleep all the time.

Techniques like these work and shouldn't be dismissed just because they feel squishier than taking a sleeping pill. But I'm putting them at number two on my list, just because I think they've probably already crossed your mind if you're a person who's been struggling with sleep.

MIGHT BE NEW: STIMULUS CONTROL

What might not have occurred to you is this: if you can't sleep, get out of bed. Even if it's the middle of the night, even if you're otherwise very comfy. Stimulus control is about breaking down and rebuilding your associations with bed so that when you see it, you just think "sleep." If you're not sleeping, get out of there.

There's the pretty straightforward part of this, which is that you shouldn't watch TV or be on your computer in bed. [54] But you also shouldn't be lying in bed in the dark, loathing yourself for not falling asleep. If it's not happening after ten minutes or so, get up and do something relaxing until you feel sleepy again. Then you can go back to bed and try again, and if it's still not happening, rinse and repeat.

A couple notes on this one: My obvious vote is to *stay in a dark or dim environment* when you get out of bed, almost as if the darkness was non-optional (like it would be without artificial lighting). I like calming audiobooks for this (very few photons coming off those audiobooks), or other audio-centric relaxation activities. The caveat here is that you shouldn't stay so in the dark that you risk tripping and hurting yourself (be careful!). Relatedly, if you try an activity with some light exposure in it, like journaling, and it works really well for you, that's great. Do that thing. The important thing isn't that you adhere perfectly to a strict circadian doctrine; the important thing is that it works.

There's also a low-effort version of stimulus control that I thought I invented but apparently plenty of other people on the internet have also discovered before:[55] if I'm not falling asleep right away in my bed in my normal orientation, I try rotating 180 degrees to see if that helps. Feet up where pillow was; pillow down where feet were. For me, this is usually different enough from the view I had in my original orientation that it breaks any negative associations I was building up by not finding sleep right away.

Going back to the CPR analogy, stimulus control is like making sure you've got proper form as you try to resuscitate someone. Lying awake in bed all night, actively building up negative associations with that space, isn't doing you any favors and is probably making things worse. It's a bit like you went in to do chest compressions for a person, missed, and ended up pushing rhythmically on their kneecap instead. Getting out of bed until you're sleepy, then, is like pausing compressions until your hands have found their way back to the neighborhood of the heart.

REFRAMING ACTUALLY WORKS: COGNITIVE RESTRUCTURING

I made up "this bed is haunted by the ghost of a Victorian child" as an example of a negative sleep-related belief that could impede a person's ability to fall asleep, which

---

54 *I know, it is great to be comfy while doing these things. This is why I bought a queen-size comforter for my couch.*

55 *But I think I'm the first person to put it in a book. If you use this and it helps you, please credit Dr. Olivia Walch (2024) for her brilliant discovery.*

is of course very silly. But there's something that particular belief has in common with the beliefs of the worst insomniacs: a sense of helplessness. "Nothing good ever comes from headin' up there," warns the creepy old man at the gas station on the way to your Airbnb on Haunted Bed Hill, and in this scenario I'm concocting right now, he's correct: whatever haunts that bed is beyond you. You're powerless, and everything's hopeless. There's no point in even trying, unless you're the one character in *Haunted Bed* who gets to live to the end.

"I can't predict sleep" or "I'll fall asleep and not be able to wake up" are feelings of helplessness expressed by some insomniacs. Or "I'll be a mess tomorrow, and it's all because I'm a failure who can't get to sleep right now."

Cognitive restructuring is all about plucking those thoughts out of your head with tweezers and replacing them with nontoxic, self-actualized alternatives. "My body knows how to fall asleep, and I can trust it," or "I've been fine before on a few hours of sleep, and I'll be fine again."

Some of my personal favorites of this flavor are:

- "Think about new parents: they sleep terribly, but our species' entire survival has been built on generation after generation of sleepy infant caretakers getting by."

- "People in pre–industrial lighting societies often don't sleep eight hours a night, and they're not stressing about it."

Even better for me is the factual knowledge that, when it comes to your circadian rhythms, you *emphatically aren't helpless*. You have the power to control your light environment, when you work out, and when you eat. If you're having a hard time falling asleep, and you get out of bed and move to another place to relax while staying in a dark, dimly lit environment, you are actively doing a great thing for your body's circadian rhythms, independent of whether you actually fall asleep or not.

Said another way: you don't have to be asleep to be doing the right thing for your circadian clock. "I'm doing everything I can to set myself up for success" is something I'll tell myself if I'm having a hard time sleeping but still staying in the dark. Or, "The fact that I'm keeping the lights low right now means I'll have an easier time falling asleep tomorrow," which is true.

The despair of "I can't fall asleep, I can't fall asleep, how is it 2:50 AM already? I can't fall asleep" is neatly sidestepped when you move from focusing on sleep to circadian rhythms. From a circadian perspective, "Am I in an environment that's pretty dark when I'm supposed to be? Yes? Cool, I'm killing it" is just about the start and end of it.

So if you're in bed, and you're not asleep, and tomorrow's the big presentation, you've got two options:

1. Punch yourself in the forehead with negative, self-defeating thoughts about how you're failing your future self by not magically becoming unconscious.

2. Pat yourself on the back for staying in the dark and practicing good sleep CPR form (getting out of bed when you're not sleeping, relaxing, etc.), the same way you might pat a horse on the nose for taking a carrot from your hand.

Cognitive restructuring is about choosing Option Number Two more.

SLEEP LESS TO SLEEP MORE: SLEEP RESTRICTION

Maybe this one isn't so surprising, given the earlier parts of this book, but it does sound counterintuitive on its face. You've got insomnia. You're desperate for sleep. The chance to sleep appears—maybe a nap, maybe sleeping in. Should you take it?

When I was in college, my answer was "definitely." I'd sleep any chance I got and plenty of times I technically didn't (lectures, whoops; tests, whoops). This made perfect sense, from the bean-counting perspective of sleep: more sleep is better, so if the chance comes up to get some, do it. Congrats; you've racked up more health points. Ignore the fact that you feel like you got hit by a truck when you wake up.

Of course, the fact that I'm even asking means the question can't be an unequivocal yes. And indeed, this part of sleep conditioning is about saying no to sleep opportunities you might otherwise have said yes to. From the two-process model of sleep, we know that a big part of sleeping well is building up a hunger for sleep and having this hunger align well with your circadian drive to sleep. If you have too little hunger for sleep (because you took a nap during the day, say), you're not going to have enough momentum to push you into sleep at night.

Remember that building up sleep pressure is like filling up a water cooler. Napping is like draining that water cooler. You might want to drain a little of the water from the cooler if you're feeling like it's putting pressure on the tap, the same way you might want to nap if you feel really, really tired. [56] But let out too much and you won't be at the level you need to be to fall asleep later.

---

[56] *And you should nap, or call off duty, if you're extremely tired but being asked to make critical decisions, drive, or otherwise operate heavy machinery.*

One of the ways sleep restriction is implemented involves limiting a person's total time in bed per day to 5.5 hours for a week or so, and then slowly increasing the amount of time they stay in bed so long as they're not having a hard time falling asleep or waking up in the middle of the night.

Here's where the CPR analogy truly starts to shine: telling yourself that you're only allowed to try to sleep from midnight to 5:30 AM, say, and that the remainder of your day will be spent awake is a form of *groove reintroduction*. You're giving your body a clear pattern to lock onto. You're pumping a rhythm back into your sleep.

You might feel some tension about cutting your sleep down to shorter than your goal at the start, which makes sense from the much-heralded perspective of "sleep health is sleep duration." But sleep rhythmicity—your regularity—matters too. And the success of CBT-I at treating insomnia shows that getting a rhythm back can pay serious dividends, even if it comes at a temporary cost to your nightly sleep hours.

NAPS

So you might not want to nap if you're a person with chronic insomnia. If you don't experience insomnia, are naps so bad?

In general, I'm pro-naps. Naps can help you recover from sleep deprivation. Naps are almost always good.

But let's talk about that "almost always." When might you want to avoid napping?

Well, maybe you're trying to shift your body's rhythms and are at a point in your biological day where getting light exposure will be very, very helpful to achieving that shift. Closing your eyes to take a nap will block photons from reaching your retinas, which means your brain won't have the photic momentum it needs to push through a shift in your rhythms. Probably not a big setback if the nap is short, but a multi-hour nap at the wrong time could end up slowing down how quickly you adjust.

Or maybe you really, really need to be alert right at the moment when you'd be waking up from a nap. In that case, you might worry about sleep inertia, the phenomenon of general grogginess and impaired performance that can persist for several hours after waking. This too might make you want to hold off on a nap.

And of course, even if you're a person *without* chronic insomnia, taking a late-in-the-day nap (and draining your hunger for sleep right before bed) probably isn't going to be great for your ability to fall asleep that night. Wanting to avoid messing up your sleep that night is another great reason not to nap.

But let's say it's the middle of the day and you're sleepy. You've got other stuff to do today. Do you take a nap? If so, how long?

Answer: you probably want a ten-minute nap.

This, like all science boiled down to a single tidbit, is a terrible simplification. It matters what your internal time is (Does your body think it's day or night?) and what your recent sleep/wake history is. And, to be clear, I'm not saying, "Get into bed at 1:05 PM and set your alarm for 1:15." You'll likely need some time to fall asleep after you get into bed, which means giving yourself a bigger window for the sleep opportunity (setting your alarm for 1:35, say).

The reason for such a short duration is that multiple studies have found a ten-minute nap during the day improves performance right off the bat, while longer naps mean that you have to sink time into *recovering* from your nap after you wake up. A twenty- or thirty-minute nap in these studies was still found to be better for fatigue than staying awake, but participants could still be shrugging off the effects of sleep inertia more than two hours after waking, while a five-minute nap was generally not enough for much of an effect.

Would you ever want to take a longer nap? You might, if your goal is not so much "perform better for the next two hours" as it is "don't fall asleep in the next ten hours." In one study from 1986, researchers kept subjects up all night, let them take a morning nap, and then measured how readily they fell asleep at different points over the rest of the day. Here, a fifteen-minute nap was barely better than no nap at slowing down how rapidly people fell back asleep, while a sixty-minute nap had alerting effects that persisted four to eight hours later. The benefits didn't keep increasing past sixty minutes, though: a two-hour nap didn't get them anything more than a sixty-minute one did.

Another reason to consider a longer nap is memory. People who get a sixty-minute nap do a better job at remembering words they've been exposed to than people who don't get a nap. That said, a six-minute nap is also enough to see a significant memory boost—for one-tenth the time investment.

In the "it's the middle of the day" scenario, you probably want a short nap. You might want a longer nap, though, if you don't need to be super alert for the next few hours, but you do need to stay awake later in the day.

Lastly, you might benefit from a nap even if you don't think of them as particularly helpful for you thanks, in part, to the fact that we all tend to walk around with a bit of sleep pressure all the time. The benefits of napping that show up in objective reaction time tests often aren't reflected in how people subjectively rate their own sleepiness. You might get more of a boost from naps than you think, and you may also need less of a nap than you'd expect to see that boost.

ALARMS AREN'T ALL BAD

I don't normally wake up with an alarm, which is a thing I usually try to avoid bringing up in polite conversation. It's a bit like saying you can eat all the salt you want with no ill health effects to a group that, statistically speaking, has some folks with high blood pressure in it. *Good for you*, think the people who do have high blood pressure, as they grit their teeth and their blood pressure rises more.

The fact that I don't sleep with an alarm might make you think I'm anti-alarm across the board, which isn't the case at all. So much of my view on sleep health is informed by keeping a consistent rhythm, and in some sense, a daily alarm is like a very, very slow metronome. The sun is also a metronome in the same way, and an alarm can act for us like the sun would when (for reasons of curtains, windows, tree cover, home layout, or work schedule) we don't get a daily dose of morning light in our face when we'd most benefit from it. For someone trying to shift their body's rhythms to be earlier, using an alarm during the adjustment period can be an invaluable tool.

Of course, there have been times when I have wanted to throw my alarm into the path of an oncoming train. Alarms, when you're sleep-deprived, can dredge up feelings of awful, bone-deep despair. These days, this most often happens to me when I'm getting up for an early flight, but for years (college) the best part of my day was getting into bed at night, and the worst part of my day was two seconds after my alarm went off, when I realized, dread washing in from all sides, that consciousness was happening again.

If you really need an alarm to wake up, or if you're in a hate-hate relationship with yours—if it goes off and keeps going off, and you sleep through it, or you snooze it eight times, or you throw it into a wall—probably, something's off. That's not an alarm acting like a helpful metronome friend. That's an alarm being an accomplice in the systematic denial of you getting enough sleep.

Just like you'd expect based on the Long Dark experiment and the camping studies (and, you know, all of this book so far), tuning your light exposure could be one way of rehabilitating your relationship with your alarm.

I've run experiments in sleep labs where we've had people come in around 5:00 or 6:00 PM to do a dim light melatonin assessment followed by polysomnography. This means keeping them in the dark for six hours to collect their spit, then plopping them into bed for what is hopefully a representative night of sleep.

"My normal bedtime's midnight," one subject told me, right as they arrived. "But I sometimes have a hard time falling asleep even then."

Oof, I thought at the time. Given the first night effect often experienced by people in sleep labs as they adjust to having someone spy on them while they're in bed, I was prepared for this participant's troubles falling asleep to be magnified.

Yet I needn't have worried. After six hours of dim red light, they were basically a puddle of sleepiness by the time we needed to get them to bed. We were more worried about them tripping over themselves on the way to their room than we were about them not being able to fall asleep.

What happened? They got *dark*-dark. Their rhythms weren't delayed by light during their phase delay region. Like the people in the Long Dark experiment, they got a longer opportunity to sleep and they took it. And the next morning, at least in theory, their leftover hunger for sleep wasn't quite as profound as it otherwise would be.

Do this enough days and, like the people in the Long Dark study—like *me*, when I started keeping to a very, very consistent sleep-wake schedule—you might end up not needing an alarm anymore.

YOUR WHOLE BODY
IS CLOCKS

LET'S GO BACK TO the suprachiasmatic nucleus (SCN), or "central pacemaker" for your body's circadian rhythms—that nifty little cluster of neurons nestled deep in your hypothalamus. Often, people call it the conductor of the circadian orchestra. It's the one cuing in melatonin production during the biological night, the same way a maestro cues in the trombone section in "The Stars and Stripes Forever."

Increasingly, though, it's become apparent that the circadian orchestra is less "professional national symphony" and more "middle school band where everyone's been allowed to keep their phones and is texting mid-song." The circadian processes in your stomach, liver, and muscles—your *peripheral clocks*—pay attention to the signals from the SCN, but they *also* pay attention to the food you eat, the drinks you drink, and the times you're active. It's not just your brain deciding what to do with the signals that you give it in an effort to figure out what time it is; it's every part of you.

This means, of course, that all the notions we've talked about for your brain and sleep (being in a groove, consistency, amplitude, etc.) apply to your other organs and other rhythms as well. Food at the wrong time should be expected to confuse your clock the same way light at the wrong time would. An intense workout at 3:00 AM will probably do more than disrupt the pacemaker in your brain—it'll likely confuse your musculoskeletal clocks as well.

There's another implication of this, though, which is that just as neurons in the SCN can become desynchronized from each other, so too can the instruments in the circadian orchestra become desynchronized if half are watching the conductor and half are messaging their pals.

As a field, circadian scientists are only just beginning to explore the implications of what *internal desynchrony* across organ systems means for health and well-being. One of the reasons for this is that we don't have super easy ways of telling what time your heart or liver or stomach thinks it is, the same way we have agreed-upon "gold standard" markers of the SCN, like DLMO. But the shape of what happens when you disrupt clocks outside the brain is starting to emerge. That's what this section is about.

BAD BRAIN TIME: CIRCADIAN RHYTHMS AND MOOD

Being, as I am these days, a circadian zealot means that I don't see the hours of 2:00 AM to 4:00 AM very much anymore. Sure, I'll sometimes wake up around then, keep the lights out, move downstairs to the couch, and then fall back asleep again, but it's not like I'm pulling all-nighters.

In my chaos days of sleep, though, they were regular companions. My very first all-nighter was in sixth grade, for an English class assignment. We were supposed to write a mini-book of our own poetry, with every page illustrated and colored. "I don't want to see any white space" is what my memory has the teacher saying. "Color the whole thing."

Coloring whole things takes FOREVER, I learned that night, as I meticulously shaded in my drawings of leaves for my poem about leaves. I colored as the clock hit 11:00 PM, then midnight, then 1:00 AM. At 2:00 AM, I started to feel terrible. It wasn't just sleepy—we've talked about the circadian drive to sleep peaking during the night a bunch already—but anxiety, distress. I was never going to be done. My coloring became loose and despairing as I filled in a gray mountain for a poem about mountains.

Around 5:45 AM, though, something changed. I started to feel good—really good. A sleep-loss induced high. I colored a beautiful sunrise on the last page as I looked out at a real one coming up over the fence in our backyard. I felt wise, like the edges between me and the universe had blurred. I was totally going to get an A.

Of course, this was my circadian clock at play. My circadian drive to sleep was peaking for me between 2:00 AM and 5:00 AM, sending those waves of sleepiness that made my coloring bad and my eyelids heavy.

But the way my mood sank and rose wasn't *just* about sleepiness. Generally speaking, the longer we stay up at night and the more our core body temperature drops, the more people tend to be impulsive, dangerous, and unable to keep their thoughts and actions in check. We have bad brains.

Here's what suicidal risk looks like as a function of time over the course of the day:

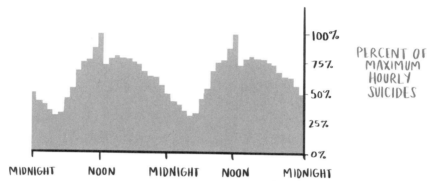

Here's what it looks like when you control for the number of people awake at each time of the day:

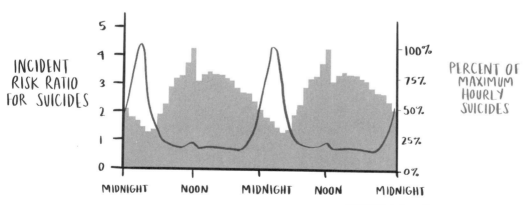

FIGURE 1 from Tubbs et al., 2022

In other words, while there are fewer people awake at 3:00 AM, the risk of suicide is much higher around that time.

We see this kind of trend in shift-working medical interns too, as well as in people undergoing constant protocols aimed at isolating their circadian rhythms. Both of these are at least somewhat stressful situations to be in, but once every 24-ish hours, in a circadian-driven way, their brains decide to make their lives much, much worse.

Losing sleep is part of it, for sure. Staying up can affect the chemicals in your brain, as well as the ways different parts of your brain talk to each other. But a number of neurotransmitters, including dopamine, follow circadian rhythms, meaning that they can go up or down at night even if you're not sleeping.

In a sense, being awake at night is a bit like being on an ice rink when a Zamboni is cleaning it. You're not supposed to be there, and normal rules don't apply. It's slippery. You can fall on your face (ice rink version) or fall into a long rumination on an embarrassing thing from 2012 (staying up late version).

Speaking of that: rumination is when you're in a repetitive, internally directed thought-spiral. You're not looking for solutions, or at least you're not finding them. You're just spinning your wheels and thinking about the same things, unproductively, over and over again. [57]

What should you do if you're up at night and you fall into a bout of rumination?

A. Solve your problems once and for all. It's time.

B. Distract yourself with something completely unrelated.

The answer, at least according to some research, is B! Distract yourself. Stop trying to think about your problems—you're not going to fix them with your terrible, up-at-night brain. Instead, think about *anything else*, the more visual the better. In one of the studies looking at strategies for rumination, participants who had been primed to feel bad [58] were told to visualize "'what a lemon looks like when you shine a bright light on it,' cutting a lemon with a knife, smelling the lemon zest/juice, and holding a lemon slice close to their eye and squeezing it." *And it worked.* Their mood improved, and their heart rate chilled out.

---

57 *Rumination is almost always negative, but it IS possible to have a positive ruminative cycle— thinking, for instance, about how awesome you are, on repeat.*

58 *By playing Cyberball. Quoth the paper: "Cyberball is a virtual ball-throwing game in which participants believe they are playing catch on the computer with two same-aged peers, who are actually pre-programmed computer players. After several rounds of play in which all players throw the ball equally to one another, the computer players throw the ball only to one another, excluding the participant."*

Similar results hold if you're having a hard time falling asleep after something stressful happened during your day: *think about something else.* In another study, students who had been given a stressful midterm test earlier in the day were split into two groups. The first group was told to "'think about how you felt when you were taking the test today,' 'think about why you had the reactions you did during the test,' [and] 'think about the possible consequences of your grade on the test.'" The second group was told to think about clouds and the layout of the local shopping center.

The students were further categorized based on a pre-trial survey as high-trait ruminators (i.e., prone to ruminating) or low-trait ruminators (not that prone to ruminating). While both the high- and low-trait ruminators were fine in the "think about some clouds" group, the high-trait ruminators told to dwell on their test before bed reported significantly worse sleep quality than both the low-trait ruminators in the same group and the high-trait ruminators who were thinking about the nearby mall.

There's a lot more to think about with regards to how mood, sleep, and circadian rhythms interact, but we should do this thinking during our biological day. If it's the middle of the night, think about cutting a lemon with a knife instead.

FOOD AS A TIMING SIGNAL

EATING AROUND THE CLOCK

In my chaotic years of not sleeping (college), I could eat pretty much any food in any amount at any time of the day. One of my grossest inventions was the "s'moretine," which was a s'more (graham cracker, chocolate, marshmallow) made out of the ingredients that were available to me in my school's cafeteria. Since they didn't have chocolate bars, I used M&M's from the ice cream bar, and since they didn't have graham crackers, I used saltines. I would cook these in the microwave and eat them at 6:45 AM in the morning.

After I stopped having a wildly erratic schedule and got my circadian act together, something changed. I just . . . didn't feel hungry after a certain point in the evening. And if I tried to eat something very late in the day (say, a midnight snack), it made me feel . . . kind of bleh.

I've never been one to hold back on eating when hungry. But once I was on a regular schedule, the nonstop, slow-burning hunger I'd had back when I was up at all hours started to dissipate. I still felt hungry, but only at certain times of the day. The rest of the time I wouldn't really feel hungry at all. You might say that my eating and hunger patterns became *more strongly rhythmic.*

There's an analogy to be made here to the feelings of "flatness" experienced by shift workers. It was like my hunger rhythm had flatlined, and keeping to a consistent light-dark schedule helped me slowly rehabilitate it.

It makes sense to think that our bodies might be more prepared to handle food at some times (like when we're awake) rather than others (like when we're supposed to be asleep). And the same way light at night confuses and disrupts the central clock in our brain, so too could food around the clock confuse and disrupt the peripheral oscillators in our organs.

In fact, this is what happens. When you move around meal timing but keep light timing fixed, glucose and insulin rhythms chase the food and leave the light signal behind. Rhythms that were previously locked together start to decouple:

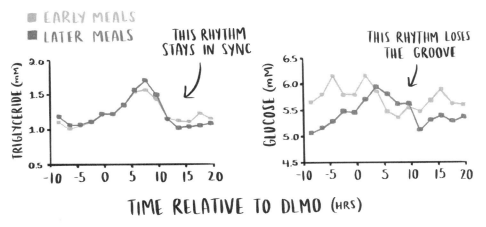

ADAPTED FROM FIGURE 2 from Wehrens et al., 2017

In the tug-of-war between light and food, who wins? It doesn't matter: you lose either way. Eating at times out of sync with your typical light and sleep has been linked to worsened blood pressure and out-of-whack protein expression. Shift workers, who often eat around the clock for reasons related to their job, have tons of metabolic issues, including higher risks of diabetes and hypertension.[59]

You can think of piping food into your digestive system at all hours of the day the same way you'd think of a newborn waking their parents up at all hours of the day: it's disruptive to your metabolic rhythms like the baby is disruptive to their

---

59 *In one survey I ran, almost 40% of shift workers said weight gain related to night shift work had made them consider leaving their job.*

sleep rhythms. "Woof, that sounds hard," I'd say to a new-parent friend of mine back in college, as she described her round-the-clock wake-up schedule. "Can't imagine what that would do to your body." That night, I would eat a s'moretine at 2:55 AM.

TIME-RESTRICTED EATING

This idea that there are times when your body is more ready or less ready for food, as coordinated with the clock in your brain, has led to the notion of "time-restricted eating," or TRE—keeping all your eating in the same window of time every day. Usually, this window is eight to ten hours long. If you get up and start eating at 8:00 AM, you might restrict your food intake to ten hours a day and stop meals after 6:00 PM. Or you might hold off on eating until 11:00 AM, in which case you'd wrap up food for the day around 7:00 to 9:00 PM. You might only do this for five days of the week, or you might do it every day.

TRE is best associated with Dr. Satchin Panda of the Salk Institute, whose group has found that simply restricting eating windows led participants in their studies to feel more energetic, sleep better, and show improvements in cholesterol and blood sugar.

I've functionally adopted TRE through the way the patterns in my cravings for food have changed. It is extremely easy to skip out on a late-night snack if your first response is, "Ugh, I'll feel terrible if I eat that now; maybe tomorrow" and not "Sounds delicious."

I attribute most of the way my eating habits have changed to more regular sleep, activity, and light boosting the amplitude of rhythms across my body, making it so I feel like I've got a clear day/night signal when it comes to food.

There's another potential explanation too: When you're awake for a bigger chunk of the day (denying yourself sleep), you burn *more* calories than you would if you were sleeping a healthier amount. But your body *overcompensates* for this increase in calories burned by making you crave more food than you need.

So another explanation for why I'm less hungry at all hours nowadays is that I'm less hungry in general. I'm less hungry because I'm getting enough sleep.

Obligatory disclaimer that not everyone can manage TRE with their job (night shift workers, for one, often have to fuel whenever they can), and further obligatory disclaimer that people with health conditions should talk to doctors before giving it a try. When it comes to anything food-related, no one solution is going to work for everybody—but for me, getting my sleep into a groove led to me inadvertently falling into an effortless food groove as well.

WHEN SHOULD YOU EAT?

Breakfast or dinner? Should you front-load or back-load your calories? If you're skipping a meal, which meal should you skip?

Here are the facts: in pretty much every head-to-head between eating earlier and eating later, eating earlier wins. Later food intake is linked to impaired glucose tolerance, reduced energy expenditure, more challenges with weight loss, and worsened outcomes after bariatric surgery.

Here's my asterisk: it really does depend on your body's biological time. Take, for instance, someone who is very adjusted to a day schedule. We know from experiments that if you put this person on a one-off night schedule, they do better *not* eating during the night as opposed to eating during the night. Is this what all shift workers should do, in general?

Probably not! After all, remember that local time isn't your biology. If I put you on a nights-forever schedule and shifted your body's rhythms so that, according to all core and peripheral clock markers, you looked exactly like someone living on the opposite side of the world, you should almost certainly eat during that side's "day," a.k.a. your local night.

So for people on the shift-work spectrum (lots of us), there's some ambiguity about best times. Probably, it's pretty well-aligned with your wake schedule, but the more irregular you are, the more unclear it becomes. [60]

Here's my take: More important than front-loading or back-loading your calories during the day is simply finding a mealtime groove—a pattern of on and off for calories like you have for light exposure and sleep.

Eat at some times (usually not too long after waking up) and avoid eating at others, the same way you'd seek light at some times and avoid it at others. Not as a hard rule (don't *not eat* if you're really hungry, the same way you wouldn't *not turn on the light* if you had an emergency in your house at 1:00 AM), but more as a general guideline. Think of craving food as the biological rhythm it is—one that, like sleep, can get stepped on, shifted, flattened . . . and also repaired.

---

[60] *Figuring out things like "best meal windows for a person on this schedule" are the types of problems we work on at Arcascope.*

FERTILITY, PRE-TERM BIRTH, AND YOUR BODY'S CLOCK

I've talked about how getting light at night squashes your body's melatonin rhythms and, in general, how light at weird times would be expected to suppress this notion of circadian amplitude. What other hormones could it affect?

Note that I said *affect* instead of suppress. You can imagine scenarios where rather than a flattened rhythm looking like this:

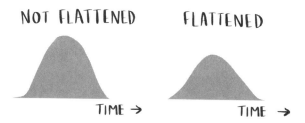

... the flattening ends up looking like what happens to cortisol as you age, where the relative difference between the highest and lowest points seems to diminish, even though both are elevated:

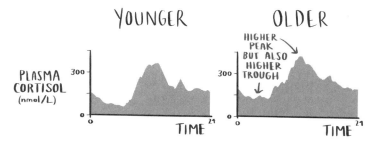

ADAPTED FROM FIGURE 1 from Van Cauter et al., 1996

You can also imagine interactions along the lines of "A suppresses B to keep its amplitude in check, but then A's own amplitude goes down, causing B's amplitude to get bigger." So there are a lot of ways losing a groove could throw off the balance of hormones in your body. This may matter if, for instance, you're trying to have a kid.

Luteinizing hormone (LH), a key hormone for fertility, is under circadian control. There are a number of rodent studies showing how messing with light exposure messes with LH. When you abruptly, temporarily jet lag some mice, their peak LH drops by 30-50% before recovering, while chronically jet-lagged mice have LH rhythms that are squished and stay squished for a long time. This, in turn, has implications for their fertility, with chronically shifted mice less than half as likely to have pups than control mice.

Mice aren't humans, but these results in mice jive with what we see in humans: that night shift workers (our real-world proxy for chronic circadian disruption) are more likely to struggle to conceive and more likely to miscarry than people not under conditions of constant circadian stress.

Sleep loss also almost certainly plays a role here, independent of circadian rhythms, though it can be difficult to disentangle circadian effects from sleep loss effects in large, retrospective studies. But focusing only on getting enough sleep, and not when that sleep is happening, means missing out on the story circadian intuition tells us: in general, a weird schedule means amplitudes get affected and phases get shifted. Circadian disruption of this kind means you might have less—or too much—of some key hormones you need when you need them.

Shift work has also been weakly linked to a higher risk of early preterm birth, as has poor sleep. Which I'm sure is great to read if you're a pregnant shift worker who just landed on this page. If that's you, let me say: not every study finds this, and it's not so much of an effect that the CDC is going around saying you shouldn't work nights if you're pregnant.

On top of that, work hours are only part of the story. You can still exert control over your light and activity during off-hours to send the most consistent, clearest day/night signal you can. One of my active research projects right now is developing a tool to help pregnant people boost their circadian rhythms, all aimed at the goal of improving outcomes and reducing pre-term birth.

THE BEST TIME TO BE GOOD AT SPORTS

Whenever the Olympics roll around, I watch the athletes compete, get all misty-eyed when they do a good job, and then stalk them on social media. *I have a shirt that looks like that*, I think as I scroll through photos of them lifting approximately three times my body weight. *We're not so different, you and I,* I think as they hit a five-inch target from a football field away.

It was in doing this during the 2020 Summer Olympics that I noticed something that caught my eye: for at least a couple of the U.S. contenders, their Instagram posts about "heading out to Tokyo" were happening . . . five days before their event.

This surprised me for the simple reason that international travel = jet lag = circadian disruption = presumably disrupted sleep and performance. And with Tokyo being thirteen to sixteen hours ahead of the U.S. time zones, I'd expect U.S.

athletes to take about six to eleven days to adjust. Of course, there are financial reasons why people can't afford to travel to Tokyo a month ahead of time to fully adjust to the new time zone, but it still had me wondering: What were they doing with their light exposure to entrain during those five days? And how much of their performance could be explained by their circadian phase at the time they were competing?

The rule of thumb for jet lag is that travelers shift about an hour per day, which means that delaying your clock by nine hours should take you about nine days. In theory, you can shift faster than that by getting light at all the right times and avoiding it at others, so that a nine-hour shift can be achieved in three to four days. But it's unclear how often travelers put schedules like this to work in the real world. Regardless, one thing Olympians should be aware of is that there are shortcuts that can help them adjust to a new time zone faster than they would if they simply tried to adopt their normal wake and bedtime routines upon landing in the new time zone.

But there's another component to this too, which is that question of "What's their circadian phase at the time they're competing?" People tend to be best at physical performance in the early evening, which aligns with a peak in core body temperature (about 10 hours before CBTmin), though it depends (of course) on their chronotype and recent circadian behavior.

Here's how peak time varies for early chronotypes, late chronotypes, and people who are sort of in the middle:

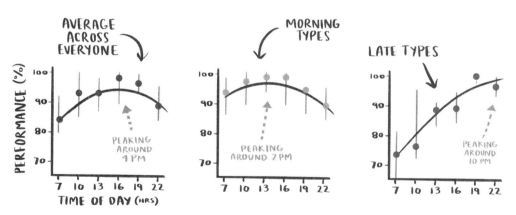

FIGURE 2 from Facer-Childs & Brandstaetter, 2015

But everything in these plots only applies if you're *entrained* to their time zone. If you transplant me into a new time zone, my body's "early evening" could be happening at 7:00 AM local time, or noon, or 3:00 AM in the morning. My body's time at that point is all in my head. [61]

In other words, there are two circadian goals you might have as an elite athlete crossing time zones. You'll probably want to entrain as quickly as possible to get over the disrupted sleep and general feelings of bleh that come from jet lag. But you might also want to entrain yourself to a schedule that aligns your body's peak performance with the time you're supposed to compete. Because if you're supposed to lift weights at 10:00 in the morning, you can *probably* give yourself an edge by tricking your body into thinking it's really later in the day.

There's data to back this idea up. In a neat 2020 paper, the authors looked at time-of-day effects in Olympic swim times from 2004 and 2016. Using local time as a proxy for internal time with the Olympics is a little tricky, in that we don't know how entrained the athletes were at the time of the competition. But it's good as a first-order approximation, since most of them had probably been on-site for at least a few days.

The researchers found that there were clear effects of time of competition on performance, with slowest swim times happening in the early morning and fastest swims in the late afternoon. These effects were big enough to exceed the difference between gold and silver in 40% of the finals, and between silver and bronze in 64% of the finals. Could some silver medalists have gotten gold just by shifting their circadian clock? This work seems to suggest that it's possible.

There's a lot of cool science to be done in this space, including *actually trying* to shift athletes in this way to see if it helps them on game day and incorporating their individual circadian state at start into the plans they follow. Teams too could come up with plans that shift their night owls more than their early birds (for example), boosting overall group performance in a tailored way.

There are some folks out there taking steps to do this already, but it can be a bit of an uphill battle to convince the broader world of elite sports. After all, we're just now getting some people to take sleep seriously—taking *light* seriously is a step beyond that.

I think we just need an Olympian to win gold and say, "Thank you, circadian rhythms" on the podium. If you think you might be this Olympian, reach out. I'm on social media. We're not so different, you and I.

---

[61] *and in my peripheral clocks*

ACTIVITY, ENTRAINMENT, AND WEIGHT GAIN IN KIDS

Here's a fact about school-age kids: they often gain weight during the summer. On the one hand, this seems surprising to me. *Shouldn't they be up and running around, building forts, and making sepia-tinted memories?* On the other hand, all I did in my childhood summers was sit in the A.C. and watch anime. So I get it.

A curious footnote to the fact that kids often gain weight in the summer is that the total amount of activity they do *isn't associated with how much weight they gain*—at least not in the data of some of my group's collaborators. A kid who really is up and running around all break might still gain weight. A kid watching *Pokémon* all day might not.

Metabolism is complicated, so I don't want to undersell how much could be going on here. Maybe one of the highly active kids was lifting all day and put on a ton of muscle weight. Maybe watching cartoons instead of being highly active reduces your cravings for food, so had my waifish childhood self been in this study, she would have had reduced caloric burn, sure, but reduced caloric input as well.

But a neat thing from some analysis we did on this dataset was that while *total* activity didn't predict weight gain, the rhythmicity of activity did. Or rather, the "entraining strength" of activity.

A kid would have strongly entraining activity patterns if they got lots of activity at the same time every day and lots of inactivity every night. They'd have weakly entraining activity patterns if they either had a low amount of total activity in general, or if they had lots of activity but it was irregular, scattered around from day to day. In the language of this book: the kids were less likely to gain weight if they were in an *activity groove* and more likely to gain weight if they weren't in one.

What does this mean? There's only so much what we can conclude from a standalone study, like this one was. But it's just another piece of evidence suggestive of the broader story: one of caring less about tallies, like total step counts in the day, and more about *when* those steps happen.

CIRCADIAN MEDICINE

WHEN YOU TAKE A PILL AND IT MAKES YOU WORSE

MELATONIN IS A HORMONE produced naturally by your body that rises and falls once a day. It is also a minimally regulated supplement you can buy at CVS and take whenever you feel like it. There's a lot to be said about the minimally regulated part of that last sentence—the melatonin you buy might not be anywhere close to the quantity advertised, and it might not even be majority *melatonin*—but I'm going to focus on the "take whenever you feel like it" piece. The fact that you can take melatonin whenever you want is important because melatonin, like light and activity, does different things to your body at different times of the day.

That's because melatonin, like light, like exercise, has a phase response curve:

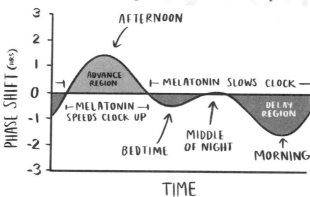

FIGURE 5 from Burgess et al., 2010

In other words, melatonin is going to shift the timing of your body's clock when you take it, in addition to any "making you sleepy" effects it has. This might be what you want; after all, if you're trying to fall asleep earlier in the day, maybe you want to speed up (or phase advance) your body's rhythms. If that's your goal, though, you want to be careful to take melatonin at the right time. Time it wrong and you could end up *slowing down* your body's clock and making it harder for you to fall asleep the next day.

Most people probably take melatonin around the time in their phase response curve where the phase shifting effects are pretty small. In other words, they take it at a time where it's not really going to speed up or slow down their body's clock. It might be able to help them fall asleep faster due to non–phase shifting effects of melatonin, but from a circadian perspective, it's a wash. If they wanted to take it to actually speed up their clock (and perhaps become more of an early bird), they'd want to take it in the advance region.

For many people, this is going to be in the middle of the afternoon—say, 4:00 or 5:00 PM. You wouldn't want to take melatonin at a time like this if you were going to be driving or otherwise operating heavy machinery afterward, but if you were heading several time zones east and speeding up your clock was the goal, and the only thing on your to-do list was "sit on plane," then sure, yeah, take that early melatonin.

But what about someone who's had to stay up one random night doing something critically important[62] and now reaches for the melatonin at 1:30 AM to help themselves fall asleep? They're probably taking that melatonin during the phase delay region.

Or what about a worker who's got a 7:00 PM to 7:00 AM shift today and a 7:00 AM to 7:00 PM shift a few days later? Taking melatonin after their shift runs a good chance of delaying their clock, when what they almost certainly want to be doing to prepare for their next shift is advancing themselves.

In both of these cases, the person's circadian clock will have shifted itself to be more like someone in a time zone west of them. This means it's more likely that they'll have a hard time falling asleep at night, at least based on the signals their circadian clock will be sending.

Pill-based melatonin, in other words, cuts both ways. It could help you achieve a goal, but it could also move you further away from that goal without you necessarily realizing it. What matters, critically, is *when* you take it.

---

62 *e.g., coloring a sixth-grade English assignment*

MORE EFFECTIVE, LESS TOXIC

Melatonin's not the only drug that's like this. The reason melatonin's timing matters is because your body is always grooving. It's changing itself over the course of the day, so the same stimulus (a pill of melatonin) is interpreted differently by your body (speed up clock at some times, slow down clock at others). Light exposure, of course, is similar: we ingest the same stimulus (looking at a screen) at different times and get different effects on our bodies.

What if instead of the pill being melatonin, which is treated like a supplement in the United States, the stakes were higher? What if it was your chemotherapy?

It's not a wild notion. For one, people have been talking about "chronotherapy" or "chronomedicine" for decades. For another, half the human genome is rhythmic. If your drug is supposed to act on a downstream byproduct of one of those genes, there's probably a best and worst time to dose with it. It's been estimated that 80% of drugs have some kind of time-of-day effect that could be exploited. But it's not (everyone groans) about the time on the clock on your wall. It's your biological, circadian time.

Here's an example of how a circadian time-of-day effect can matter in something like cancer:

Glioblastoma is a terrible brain cancer where your expected survival after diagnosis is usually only around fifteen months. There's a pill-based treatment for glioblastoma called temozolomide (TMZ), which was heralded as a major advancement for the field when it was found to extend life by two months.

Temozolomide works by entering a tumor cell and sticking a methyl group to its DNA, which eventually leads the tumor cell to die.

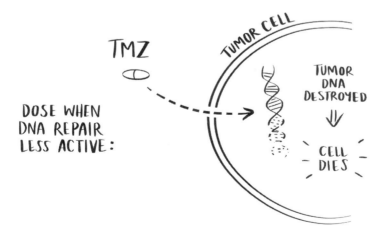

TMZ

TUMOR CELL

DOSE WHEN
DNA REPAIR
LESS ACTIVE:

TUMOR
DNA
DESTROYED

CELL
DIES

That is unless the tumor has O6-methylguanine-DNA methyltransferase (MGMT) around. MGMT is a DNA fixer. If MGMT sees the damaged tumor DNA floating around, it will patch it right up, and the tumor cell will live to see another day.

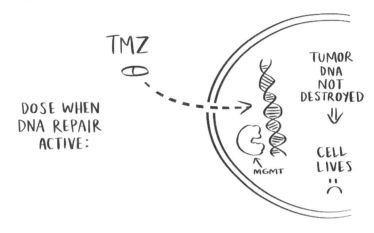

It's been known for a while that people who have MGMT floating around in their tumor cells do worse on treatment than people whose MGMT has been silenced (who don't have to combat the self-healing benefits MGMT can bring). But it had *also* been known for a while that even for people with MGMT, not all dosing times were equally effective. Dosing with TMZ at certain times seemed to shrink tumor size three times more than dosing at others.

In 2023, researchers figured out why: Tumor cells of this kind of cancer still have their circadian rhythms, which means their MGMT expression *also* still exhibits a rhythm. And dosing with TMZ when MGMT was near its peak could squash the drug's efficacy. After all, any methyl groups added to the tumor's DNA by the TMZ could be swiftly removed when MGMT was high, vs. dosing with TMZ when MGMT was nearing its lowest point and wasn't really around to remove anything.

Based on this, the "right" time to dose with TMZ to maximize efficacy for most people would be early morning, and indeed, when you look at retrospective data, the people who do the best on TMZ are those who are told to take it in the morning. Not only that: the effect size of AM vs. PM dosing is close to two months more of overall survival, which is essentially the same as the effect size TMZ was approved for in the first place.

Since the original drug trial delivered TMZ mostly in the morning, it's possible they got lucky, from a timing perspective. If they'd given TMZ mostly in the evening in the original trial, they might not have seen much of an effect, and TMZ might never have been approved.

Biology is complicated, and this same theoretical approach likely can't be copy-pasted onto other cancers. But it's a simple story: something's going up and down in your body. *Use that information.* Strike when the tumor is weak; dose when the pill has the biggest effect. The effect size you see might not be a 5 or 10% improvement. The effect size could be everything.

SHIFTING YOURSELF

Temozolomide is an interesting case because it's a chemotherapy you can take whenever. You can have it whenever you feel like it; the pills are just sitting there in your home. From this perspective, there are almost no barriers to time-optimizing it. If you decide you want to take it at 5:00 AM, you can.

Not true for other treatments. If you're going into a hospital for immunotherapy, or some other kind of infusion, or *surgery*, then good luck scheduling that for whenever you'd prefer. You have to work within the constraints of the hospital's scheduling system and the needs of all the other patients who *also* need to get scheduled. If your best time for an infusion is 2:00 AM, you're not going to get it.

One article, in discussing how heart surgery times (which seem to be best in the afternoon) could be better aligned with a person's circadian rhythms, quotes a pharmacologist who imagines "a drug that artificially 'jet lags' a patient's heart into thinking a morning surgery is actually happening in the afternoon."

Don't get me wrong: I know people love it when the solution to a problem is "take this single pill and then don't worry about it ever again." But we already do have a (yeah, okay, relatively slow-acting) drug that can do this, with an incredible track record: light exposure. *Time is just made up.* You can shift yourself with light exposure to be in whatever time zone you want and make it so the "best" time for you lines up with the "best" time for your surgery.

Will people actually do this in practice? For lots of people, no. Getting patients to straight-up take their pills *at all* is already a real challenge in medicine; getting them to substantially change their behavior in the week or weeks before a scheduled appointment will be a nonstarter for some. Plus, for others, their life and work constraints will make it harder for them to stay on an earlier or later schedule in a stable way (think: family clanking downstairs around at 7:00 PM while you're trying to sleep).

But the fact that some people won't want or be able to shift themselves to a better "time zone" prior to an infusion or surgery doesn't mean *nobody* will. And if the circadian effect sizes are large as have been reported for some treatments, that could motivate more people to take the leap.

JUST HAVING A BETTER CIRCADIAN CLOCK IS GOOD TOO

We've talked already about how getting into a groove is not about having *just* bright light exposure or *just* true darkness, but instead having a clear, consistent signal that screams "THIS IS DAY" to your body at some times of the day and "THIS IS NIGHT" to your body at others.

Being in constant total darkness leads to free running, where you fail to lock onto a 24-hour day and drift forward at your body's intrinsic period, completely decoupled from whatever the sun is doing outside the reaches of your senses. Being in constant light does basically the same thing, except it's way more miserable.

How could it not be? Constant light exposure sounds like torture. Imagine never getting a break from some level of brightness, struggling trying to sleep in a too-lit space, waking up feeling exposed and without any sense of where you are in space and time.

This is roughly what it's like for people in the hospital. Instead of constant darkness or constant bright light, though, they get a constant level of *bleh* light: too much in the nighttime, too little during the day.

Oh, and remember: the clock in your brain that listens to light exposure isn't the only clock in your body. You've got clocks in your liver, your stomach, your entire metabolic system. So just like how sending round-the-clock light to your brain is bad, so too is sending round-the-clock "food" to your food-processing clocks.

Unfortunately, this is exactly the standard of care. Total parenteral nutrition, or TPN, is a type of nutrition from an IV drip that skips the digestive tract. Getting well enough to get off TPN is often seen as the key precursor for a patient to go home from the hospital. But TPN is typically given either only at night or all the time.

This is akin to giving someone their meals either only at night after keeping them in the light all day, or *constantly feeding them*—neither of which sends a nice, clear entraining signal that communicates "DAY" at some times and "NIGHT" at others. It's about as bad of a way as you could think of to deliver TPN, thinking purely from a circadian perspective, and it's the way it's done at hospitals around the world.

Researchers at Cincinnati Children's Hospital tried a very straightforward idea: what if they restricted TPN only to daytime hours in their pediatric population, instead of making it a constant infusion? Would that improve outcomes?

It sure did: Not only did the kids on daytime-only TPN maintain as healthy a weight as the kids on TPN—they were also able to switch off of TPN entirely four days earlier. This, in turn, meant they could be ready to go home sooner, saving significant money in healthcare costs.

If CBT-I was "sleep CPR"—a way of reintroducing a rhythm to your sleep—then daytime-only TPN can be thought of as CPR for your metabolic rhythms, with the added dimension of wanting the day signal for your metabolism to align with the day signal in your SCN. And this is likely just the beginning. While biology is always complicated and progress in medicine is often slow, it sure seems like the more we do to reduce loss of rhythms and internal desynchrony in patients, the better outcomes will be.

Hospitals right now are set up for continuous, round-the-clock care, which is great. I don't want to be turned away from care just because I showed up at 2:00 AM. But the future of hospitals *has* to include more consideration for the biological rhythms of its patients and workers.

And speaking of futures . . .

A CIRCADIAN FUTURE

I STARTED THIS BOOK WITH eight hours a night—the "eight hours of sleep a night is an optimal sleep duration goal" health tidbit, which you'd probably heard about before ever cracking it open. What I hope happens in the future is that *rhythmicity of activity* and *regularity of sleep* become such commonplace, standard ideas that someone reading this book will get to this part and be like, "Yeah, duh. Nothing new here."

Here's how I think we get there.

QUIT IT WITH THE NO-SLEEP BRAGGING ALREADY

The dark truth of my college days is that I wasn't *just* not sleeping. I was not sleeping, and I was bragging about it. I'd tell people I hadn't slept in thirty-six hours and expect them to be impressed, as if staying awake took grit and character and not just a big bag of Cheerios I could eat, one by one, hunched over my dorm room desk.

If I were to be put into a room with my twenty-year-old self today and she were to start talking about her sleep habits, the first thing I would think is, *Aw, look at her. She's adorable. She'll be even more adorable once she figures out how good she looks with bangs.* The second thing I'd think is, *Wow, she is garbage at time management.*

Because that's all it is, at the end of the day. Short of mandatory work hours and life constraints that really, truly don't give you enough hours in the day, [63] going profoundly without sleep is a sign that you're trying to fit too much in.

Saying "I can operate on four hours of sleep" is a bit like saying "I can operate my phone on 2% battery." Sure, good for you. So can lots of other people. That doesn't make you a wizard of phone usage, nor is it particularly impressive that you're okay with using a laggier, worse-functioning phone. The correct response to your friend's phone running out of battery isn't, "Wow, what an incredible sign of inner strength and go-get-'em moxie!" It's, "Uh, okay. Sounds like you should charge it next time."

I think this has already been much improved in the time I've been alive, but I still run into no-sleep bragging every once in a while. Which, c'mon, people. I'm sure you have something actually impressive to talk about.

BETTER LIGHTING AT SCHOOL, WORK, AND IN THE HOSPITAL

Every year or so there's a study that's like, "We made light brighter in this school and/or work environment, and everyone was happier and smarter and more productive." And then nobody does anything about it.

Or rather, progress toward doing something about it is slow. To paraphrase something I heard once from an executive at a lighting company, "We tried for thirty years to make money with circadian lighting products, and it never worked." They ran study after study showing improvements to learning and workplace satisfaction, and at the end of the day, the test subjects were happier and no customers bit.

There are, I think, many reasons for this. One is that the sun is free, which often makes people reluctant to install a $10,000 lighting fixture in their office building that exists solely to try and imitate the sun. It's a bit like not wanting to pay $10,000 for a nice air filter in an old, dusty building because there's fresh air for free outside. The fact that the free version exists in abundance nearby makes a hesitation to spend big bucks understandable, but it doesn't change the fact that the people stuck indoors all day aren't getting any of it.

There's also the natural constraints of budget and bandwidth. Structural changes to promote better lighting can be hugely expensive, and they have to compete with all the other things that are also important and also take money to achieve.

---

63 *If this is you, so be it. Go in peace, friend. All I'm saying is that Past Me would have said she couldn't drop a single thing from her schedule, and Past Me was lying to herself.*

Which leads me to my personal theory: when push comes to shove, people just don't really "believe" in light as a meaningful factor for health. It doesn't register the way a pill does, even when its effects are many times stronger than the pill's, as is the case with light and melatonin for shifting your body's clock.

I think this is changeable at a societal level. At night, when the light in my house is on overhead for too long, it's like there's a car alarm going off down the street from me, somehow getting louder and louder as night gets later and later. It's annoying, actively. I have to get up and turn it off.

I wasn't like this back in the day, but I'm like it now. And I think enough people infected with the idea of light as a drug could lead to big changes in how we think about our homes and offices as agents for health. We'll walk past a house in our neighborhood with harsh, bright overhead lights coming through their windows after dark and think, *Oh, wow,* that's no good, the same way we'd feel seeing a suspicious wire dangling ominously from the power line by their driveway.

We'll awaken the part of our brain that (perhaps innately) knows to reject light at the times when it doesn't want it. Then we'll make our environments more tuned to our biological needs and be happier, feel healthier, and sleep better as a result.

HOW WE COULD DO BETTER BY SHIFT WORKERS

Shift workers very rarely work jobs that are just for fun. All-day, 24-hour operations are usually operating that way because they provide critical infrastructure or emergency services, and not because someone on staff thought it might be a neat thing to try. I might need to go to the hospital at 3:00 AM; I won't ever need to buy a duvet cover at 3:00 AM.

So if we want to help people providing some of the most important services out there, we should start by acknowledging shift work's health hazards as a serious occupational risk, and not something to be hand-waved away with a "They choose to work at that time for better pay" or an "Everyone has to do their time on nights when they're just starting out."

Ergonomics and human factors groups are just now beginning to think about work hours the same way they think about back supports and repetitive strain injuries, but we need more of it. We also need training to help shift workers prepare for the demands of their jobs and work schedules that take into account the physical realities of when human bodies can and cannot sleep, vs. rules of thumb that mandate a set number of hours off, irrespective of when those hours occur.

CIRCADIAN MEDICINE, FOR REAL

We do have circadian medicine, a little bit. Increasingly, medications are coming with recommendations to "take after waking" or "take before bed," both of which are much better than "take in the AM" or "take in the PM" (*see* "Time is fake" for why). That said, circadian effects are still barely tracked in clinical trials, which means that drugs are often approved based on data that collapses all variation over time into a single number.

So long as we don't think about time—and not just wall-clock time, *biological time*—in all aspects of health and medicine, our understanding will be woefully sparse, like trying to piece together the plot of a movie with only thirty seconds of every ten-minute chunk. And it's not just medicine—we should be thinking of circadian rhythms across the whole spectrum of biological research to make sure we're not missing something critically important. We need to treat timing like the integral part of the picture it is.

HOW I SLEEP

When I first got on the sleep groove train, I was rigid: fixed times for lights on, fixed times for lights off. No exceptions. Researchers were literally going to look at my watch data, and they would know if I was lying to them about when I was in and out of bed, and I would be embarrassed.

These days, I do things a bit more by feel. I feel a certain time of the day when light seems too bright, so I turn it down. I feel like it's time to bust open the windows in the morning, so I do. Often, if I wake up before dawn, I enjoy a little bit of time in the dark before I turn on anything overhead—especially if the sun hasn't come up yet.

I don't always sleep through the night. I don't always fall asleep right away. Probably, if you were to look at what a wearable told you about my sleep, it'd say I was getting six and a half to seven hours of sleep a night. I look at my phone (with dark mode on) before bed most nights. My wake time is more regular than my bedtime. I'm not an extreme outlier in any of this.

Here's where I am an outlier:

- There's no difference between my weekdays and weekends
- I call it pretty early on nights out with my friends
- If I stay up late, I do it in the darkest dark I can find
- I will turn out lights at a friend's house to achieve this

And maybe, most of all, I don't *really* think of getting a good sleep. I think about getting "a good dark."

Was I in a low-stress dark environment for enough time last night, setting myself up for circadian success today? Can I blast myself with light today to help boost my momentum into sleep tonight? Cool. Then I feel pretty good. The sleep matters still, of course, but since it's not something I can consciously, directly control, why stress about it? Did I do everything I could do? That's enough.

SLEEPING BETTER

Often, trying to change yourself through aggressive, head-on action to prove you're serious about personal growth is like insisting on making eye contact while fighting Medusa to prove you're not intimidated. I've never been able to get myself to go to the gym more by self-berating about health or by fixating on the challenge. It's always been something structural—the friend you make plans to meet up with there, the Zumba teacher you say "See you next week!" to—that reshapes gravity so that *not going* becomes harder than *going*. The way to do it is obliquely, tweaking the social mesh and your environment to make it so that the change flows downhill, effortlessly.

Sleep's the same way. Think of your sleep as a plant growing in the garden. You can talk to your plant all you want and explain to it the myriad reasons why you, personally, feel it should consider growing, but if you burn yourself out doing this and start to despair, your friends and neighbors are going to wonder why you didn't just water it.

Cultivating a healthier circadian clock is a way of setting the environment for sleep change without looking it directly in the face. It's finding good soil for your plant and plenty of sunlight. Trust that if you do that much, the plant can take it from there.

ACKNOWLEDGMENTS

A FEW OF THE chapters here are adapted from blog posts I've written on my company website. You can read more there at https://www.arcascope.com.

A full list of citations—and any future corrections, as the science updates—is at http://sleep-groove.com/

Many people either read early drafts of this book or had conversations with me that shaped my thinking on what's inside it. Massive thanks to Philip Cheng, Jenny Moreno, Ellen Stothard, Janis Anderson, Renske Lok, Jesse Cook, Doug Marttila, Eric Canton, Franco Tavella, Tamara Brady, Arman Soudi, Rebecca Plotnick, Andrew Tsyaston, Megan Danielle, Enzo Santos, Walter Hickey, Paul Cammer, Erik Herzog, Maria Gonzalez-Aponte, Jamie Zeitzer, Andrew Phillips, Marc Ruben, Yitong Huang, Sharmin Ghaznavi, Beth Klerman, Till Roenneberg, Cathy Goldstein, and Danny Forger. If anything's wrong in this, it's my fault, not theirs.

GLOSSARY

ACTIGRAPHY: measuring activity and rest patterns with a watch that tracks motion (and often light) to infer if someone was asleep or awake

BIOLOGICAL TIME: the collective state of all the circadian quantities in your body; often simplified to a distance from a circadian marker, e.g., six hours after DLMO

CHRONIC SLEEP RESTRICTION: schedules that limit sleep duration over multiple days

CHRONOTYPE: often used to capture if you're an early bird or a night owl; for the purposes of this book, chronotype is the midpoint of your sleep on days you're not working. Can be influenced by both genetic and environmental factors and can change over time.

CIRCADIAN DISRUPTION: the state of having circadian rhythms that are out of sync or squashed relative to your ideal rhythms; the loss of a groove

CORE BODY TEMPERATURE MINIMUM (CBTMIN): the time when, if we put you on a constant routine protocol and didn't let you sleep or move, your body temperature would hit its lowest point

DIM LIGHT MELATONIN ONSET (DLMO): the time your body would start to produce melatonin above a threshold if we put you in (very dark) darkness

ENTRAINMENT: the act of adjusting to a new schedule; falling into a groove

INTRINSIC PERIOD: your body clock's "natural gait," or how fast your circadian day would be under constant darkness conditions

INTRINSICALLY PHOTOSENSITIVE RETINAL GANGLION CELLS (IPRGCS): cells in your eyes that communicate light information directly to the SCN and other parts of your subconscious visual system

LUX: a unit of illuminance that captures how bright the light is

MASKING: the phenomenon by which behavioral or environmental factors hide or otherwise obscure the measurement of circadian rhythms

PERIPHERAL CLOCKS: circadian clocks in the organs outside your brain

PHASE ADVANCE REGION: parts of your circadian day when light exposure or other timing signals speed your clock up and make it run faster

PHASE DELAY REGION: parts of your circadian day when light exposure or other timing signals slow your clock down and make it run more sluggishly

PHASE RESPONSE CURVES: plots that capture how your circadian rhythms either speed up or slow down in response to stimuli at different times

POLYSOMNOGRAPHY: the gold standard method of tracking sleep via electrodes on your head, among other sensors. Typically takes place in a sleep lab

RODS AND CONES: cells in your eyes that communicate light information primarily to your image-forming (conscious) visual system

SLEEP DURATION: how long you spend asleep

SLEEP EFFICIENCY: the amount of time you spend asleep divided by the amount of time you spend trying to sleep

SLEEP INERTIA: the grogginess felt on waking; usually dissipates within half an hour

SLEEP REGULARITY: can be defined multiple ways, roughly captures how consistent your bedtimes and wake times are

SUBCONSCIOUS VISUAL SYSTEM: non-image forms of vision, like those that feed into pupil reflex and circadian rhythms

SUPRACHIASMATIC NUCLEUS (SCN): the cluster of neurons in your hypothalamus that act as the central pacemaker for your body's rhythms

TIMING SIGNAL: signals that send timing cues to your circadian rhythms

TWO-PROCESS MODEL OF SLEEP: a framework for thinking about sleep that includes a circadian component driven by circadian rhythms, and a homeostatic component driven by recent patterns of sleep and wakefulness

WAKE AFTER SLEEP ONSET (WASO): how much time you spent awake between falling asleep and waking up to start your day

SELECTED REFERENCES

Bano-Otalora, B., Martial, F., Harding, C., Bechtold, D. A., Allen, A. E., Brown, T. M., Belle, M. D. C., & Lucas, R. J. (2021). Bright daytime light enhances circadian amplitude in a diurnal mammal. *Proceedings of the National Academy of Sciences of the United States of America*, *118*(22). https://doi. org/10.1073/pnas.2100094118

Brooks, A., & Lack, L. (2006). A brief afternoon nap following nocturnal sleep restriction: which nap duration is most recuperative? *Sleep*, *29*(6), 831–840.

Burgess, H. J., Revell, V. L., Molina, T. A., & Eastman, C. I. (2010). Human phase response curves to three days of daily melatonin: 0.5 mg vs. 3.0 mg. *The Journal of Clinical Endocrinology and Metabolism*, *95*(7), 3325–3331.

Buysse, D. J. (2014). Sleep health: can we define it? Does it matter? *Sleep*, *37*(1), 9–17.

Cain, S. W., McGlashan, E. M., Vidafar, P., Mustafovska, J., Curran, S. P. N., Wang, X., Mohamed, A., Kalavally, V., & Phillips, A. J. K. (2020). Evening home lighting adversely impacts the circadian system and sleep. *Scientific Reports*, *10*(1), 19110.

Chang, A.-M., Scheer, F. A. J. L., & Czeisler, C. A. (2011). The human circadian system adapts to prior photic history. *The Journal of Physiology*, *589*(Pt 5), 1095–1102.

Chellappa, S. L., Qian, J., Vujovic, N., Morris, C. J., Nedeltcheva, A., Nguyen, H., Rahman, N., Heng, S. W., Kelly, L., Kerlin-Monteiro, K., Srivastav, S., Wang, W., Aeschbach, D., Czeisler, C. A., Shea, S. A., Adler, G. K., Garaulet, M., & Scheer, F. A. J. L. (2021). Daytime eating prevents internal circadian misalignment and glucose intolerance in night work. *Science Advances*, 7(49), eabg9910.

Cohen, S., Doyle, W. J., Alper, C. M., Janicki-Deverts, D., & Turner, R. B. (2009). Sleep habits and susceptibility to the common cold. *Archives of Internal Medicine*, *169*(1), 62–67.

Damato, A. R., Luo, J., Katumba, R. G. N., Talcott, G. R., Rubin, J. B., Herzog, E. D., & Campian, J. L. (2021). Temozolomide chronotherapy in patients with glioblastoma: a retrospective single-institute study. *Neuro-Oncology Advances*, *3*(1), vdab041.

Damiola, F., Le Minh, N., Preitner, N., Kornmann, B., Fleury-Olela, F., & Schibler, U. (2000). Restricted feeding uncouples circadian oscillators in peripheral tissues from the central pacemaker in the suprachiasmatic nucleus. *Genes & Development*, *14*(23), 2950–2961.

Dawson, D., & Reid, K. (1997). Fatigue, alcohol and performance impairment. *Nature*, *388*(6639), 235.

de la Iglesia, H. O., Fernández-Duque, E., Golombek, D. A., Lanza, N., Duffy, J. F., Czeisler, C. A., & Valeggia, C. R. (2015). Access to electric light is associated with shorter sleep duration in a traditionally hunter-gatherer community. *Journal of Biological Rhythms*, *30*(4), 342–350.

Depner, C. M., Melanson, E. L., Eckel, R. H., Snell-Bergeon, J. K., Perreault, L., Bergman, B. C., Higgins, J. A., Guerin, M. K., Stothard, E. R., Morton, S. J., & Wright, K. P., Jr. (2019). Ad libitum weekend recovery sleep fails to prevent metabolic dysregulation during a repeating pattern of insufficient sleep and weekend recovery sleep. *Current Biology: CB, 29*(6), 957–967.e4.

Depner, C. M., Melanson, E. L., McHill, A. W., & Wright, K. P., Jr. (2018). Mistimed food intake and sleep alters 24-hour time-of-day patterns of the human plasma proteome. *Proceedings of the National Academy of Sciences of the United States of America, 115*(23), E5390–E5399.

Duffy, J. F., Willson, H. J., Wang, W., & Czeisler, C. A. (2009). Healthy older adults better tolerate sleep deprivation than young adults. *Journal of the American Geriatrics Society, 57*(7), 1245–1251.

Facer-Childs, Elise, and Roland Brandstaetter. 2015. The impact of circadian phenotype and time since awakening on diurnal performance in athletes." *Current Biology: CB* 25 (4): 518–22.

Goldstein, C. A., & Smith, Y. R. (2016). Sleep, circadian rhythms, and fertility. *Current Sleep Medicine Reports, 2*(4), 206–217.

Gonzalez-Aponte, M. F., Damato, A. R., Trebucq, L. L., Simon, T., Cárdenas-García, S. P., Cho, K., Patti, G. J., Golombek, D. A., Chiesa, J. J., & Herzog, E. D. (2023). Circadian regulation of MGMT expression and promoter methylation underlies daily rhythms in TMZ sensitivity in glioblastoma. *bioRxiv : The Preprint Server for Biology*. https://doi.org/10.1101/2023.09.13.557630

Gooley, J. J., Lu, J., Chou, T. C., Scammell, T. E., & Saper, C. B. (2001). Melanopsin in cells of origin of the retinohypothalamic tract. *Nature Neuroscience, 4*(12), 1165.

Graham, D. M., & Wong, K. Y. (2016). *Melanopsin-expressing, Intrinsically Photosensitive Retinal Ganglion Cells (ipRGCs)*. University of Utah Health Sciences Center.

Hébert, M., Martin, S. K., Lee, C., & Eastman, C. I. (2002). The effects of prior light history on the suppression of melatonin by light in humans. *Journal of Pineal Research, 33*(4), 198–203.

Jacobs, G. D., Pace-Schott, E. F., Stickgold, R., & Otto, M. W. (2004). Cognitive behavior therapy and pharmacotherapy for insomnia: a randomized controlled trial and direct comparison. *Archives of Internal Medicine, 164*(17), 1888–1896.

Jewett, M. E., Kronauer, R. E., & Czeisler, C. A. (1994). Phase-amplitude resetting of the human circadian pacemaker via bright light: a further analysis. *Journal of Biological Rhythms, 9*(3-4), 295–314.

Katz, S. E., & Landis, C. (1935). Psychologic and physiologic phenomena during a prolonged vigil. *Archives of Neurology & Psychiatry, 34*, 307–317.

Keeler, C. E. (1927). Iris movements in blind mice. *American Journal of Physiology-Legacy Content, 81*(1), 107–112.

Klerman, E. B., Barbato, G., Czeisler, C. A., & Wehr, T. A. (2021). Can people sleep too much? Effects of extended sleep opportunity on sleep duration and timing. *Frontiers in Physiology, 12*, 792942.

Lauderdale, Diane S., Kristen L. Knutson, Lijing L. Yan, Kiang Liu, and Paul J. Rathouz. 2008. Sleep duration: How well do self-reports reflect objective measures? The CARDIA sleep study. *Epidemiology* 19 (6): 838.

Lauderdale, Diane S., L. Philip Schumm, Lianne M. Kurina, Martha McClintock, Ronald A. Thisted, Jen-Hao Chen, and Linda Waite. 2014. Assessment of sleep in the national social life, health, and aging project. *The Journals of Gerontology. Series B, Psychological Sciences and Social Sciences* 69 (Suppl_2): S125–33.

Lahl, O., Wispel, C., Willigens, B., & Pietrowsky, R. (2008). An ultra short episode of sleep is sufficient to promote declarative memory performance. *Journal of Sleep Research*, 17(1), 3–10.

Lawrence, H. R., Siegle, G. J., & Schwartz-Mette, R. A. (2023). Reimagining rumination? The unique role of mental imagery in adolescents' affective and physiological response to rumination and distraction. *Journal of Affective Disorders*, 329, 460–469.

Lucas, R. J., Freedman, M. S., Muñoz, M., Garcia-Fernández, J. M., & Foster, R. G. (1999). Regulation of the mammalian pineal by non-rod, non-cone, ocular photoreceptors. *Science*, 284(5413), 505–507.

Mason, I. C., Grimaldi, D., Reid, K. J., Warlick, C. D., Malkani, R. G., Abbott, S. M., & Zee, P. C. (2022). Light exposure during sleep impairs cardiometabolic function. *Proceedings of the National Academy of Sciences of the United States of America*, 119(12), e2113290119.

Mure, L. S. (2021). Intrinsically photosensitive retinal ganglion cells of the human retina. *Frontiers in Neurology*, 12, 636330.

Pasnau, R. O., Naitoh, P., Stier, S., & Kollar, E. J. (1968). The psychological effects of 205 hours of sleep deprivation. *Archives of General Psychiatry*, 18(4), 496–505.

Peeples, L. (2018). Medicine's secret ingredient - it's in the timing. *Nature*, 556(7701), 290–292.

Phillips, A. J. K., Vidafar, P., Burns, A. C., McGlashan, E. M., Anderson, C., Rajaratnam, S. M. W., Lockley, S. W., & Cain, S. W. (2019). High sensitivity and interindividual variability in the response of the human circadian system to evening light. *Proceedings of the National Academy of Sciences of the United States of America*, 116(24), 12019–12024.

Provencio, I., Jiang, G., De Grip, W. J., Hayes, W. P., & Rollag, M. D. (1998). Melanopsin: An opsin in melanophores, brain, and eye. *Proceedings of the National Academy of Sciences of the United States of America*, 95(1), 340–345.

Rechtschaffen, A., & Bergmann, B. M. (1995). Sleep deprivation in the rat by the disk-over-water method. *Behavioural Brain Research*, 69(1-2), 55–63.

Roenneberg, T., Kuehnle, T., Juda, M., Kantermann, T., Allebrandt, K., Gordijn, M., & Merrow, M. (2007). Epidemiology of the human circadian clock. *Sleep Medicine Reviews*, 11(6), 429–438.

Roenneberg, T., Kumar, C. J., & Merrow, M. (2007). The human circadian clock entrains to sun time. *Current Biology: CB*, 17(2), R44–R45.

Roenneberg, T., Pilz, L. K., Zerbini, G., & Winnebeck, E. C. (2019). Chronotype and social jetlag: A (self-) critical review. *Biology*, 8(3). https://doi.org/10.3390/biology8030054

Rossman, J. (2019). Cognitive-behavioral therapy for insomnia: An effective and underutilized treatment for insomnia. *American Journal of Lifestyle Medicine*, 13(6), 544–547.

Slat, E. A., Sponagel, J., Marpegan, L., Simon, T., Kfoury, N., Kim, A., Binz, A., Herzog, E. D., & Rubin, J. B. (2017). Cell-intrinsic, BMAL1-dependent circadian regulation of temozolomide sensitivity in glioblastoma. *Journal of Biological Rhythms*, 32(2), 121–129.

Smit, A. N., Broesch, T., Siegel, J. M., & Mistlberger, R. E. (2019). Sleep timing and duration in indigenous villages with and without electric lighting on Tanna Island, Vanuatu. *Scientific Reports*, *9*(1), 17278.

Spiegel, K., Leproult, R., & Van Cauter, E. (1999). Impact of sleep debt on metabolic and endocrine function. *The Lancet*, *354*(9188), 1435–1439.

Spiegel, K., Tasali, E., Penev, P., & Van Cauter, E. (2004). Brief communication: Sleep curtailment in healthy young men is associated with decreased leptin levels, elevated ghrelin levels, and increased hunger and appetite. *Annals of Internal Medicine*, *141*(11), 846–850.

Treynor, W., Gonzalez, R., & Nolen-Hoeksema, S. (2003). Rumination reconsidered: A psychometric analysis. *Cognitive Therapy and Research*, *27*(3), 247–259.

Tubbs, Andrew S., Fabian-Xosé Fernandez, Michael A. Grandner, Michael L. Perlis, and Elizabeth B. Klerman. 2022. The mind after midnight: nocturnal wakefulness, behavioral dysregulation, and psychopathology. *Frontiers in Network Physiology* 1 (March):830338.

Vaccaro, A., Kaplan Dor, Y., Nambara, K., Pollina, E. A., Lin, C., Greenberg, M. E., & Rogulja, D. (2020). Sleep loss can cause death through accumulation of reactive oxygen species in the gut. *Cell*, *181*(6), 1307–1328.e15.

Van Cauter, E., Leproult, R., & Kupfer, D. J. (1996). Effects of gender and age on the levels and circadian rhythmicity of plasma cortisol. *The Journal of Clinical Endocrinology and Metabolism*, *81*(7), 2468–2473.

Van Dongen, H. P. A., Maislin, G., Mullington, J. M., & Dinges, D. F. (2003). The cumulative cost of additional wakefulness: dose-response effects on neurobehavioral functions and sleep physiology from chronic sleep restriction and total sleep deprivation. *Sleep*, *26*(2), 117–126.

Vartanian, G. V., Li, B. Y., Chervenak, A. P., Walch, O. J., Pack, W., Ala-Laurila, P., & Wong, K. Y. (2015). Melatonin suppression by light in humans is more sensitive than previously reported. *Journal of Biological Rhythms*, *30*(4), 351–354.

Wallace, M. L., Buysse, D. J., Redline, S., Stone, K. L., Ensrud, K., Leng, Y., Ancoli-Israel, S., & Hall, M. H. (2019). Multidimensional sleep and mortality in older adults: A machine-learning comparison with other risk factors. *The Journals of Gerontology. Series A, Biological Sciences and Medical Sciences*, *74*(12), 1903–1909.

Wallace, M. L., Yu, L., Buysse, D. J., Stone, K. L., Redline, S., Smagula, S. F., Stefanick, M. L., Kritz-Silverstein, D., & Hall, M. H. (2021). Multidimensional sleep health domains in older men and women: an actigraphy factor analysis. *Sleep*, *44*(2). https://doi.org/10.1093/sleep/zsaa181

Wang, Y., Huang, W., O'Neil, A., Lan, Y., Aune, D., Wang, W., Yu, C., & Chen, X. (2020). Association between sleep duration and mortality risk among adults with type 2 diabetes: a prospective cohort study. *Diabetologia*, *63*(11), 2292–2304.

Wang, Y. M., Taggart, C. B., Huber, J. F., Davies, S. M., Smith, D. F., Hogenesch, J. B., & Dandoy, C. E. (2023). Daytime-restricted parenteral feeding is associated with earlier oral intake in children following stem cell transplant. *The Journal of Clinical Investigation*, *133*(4). https://doi.org/10.1172/JCI167275

Wehrens, S. M. T., Christou, S., Isherwood, C., Middleton, B., Gibbs, M. A., Archer, S. N., Skene, D. J., & Johnston, J. D. (2017). Meal timing regulates the human circadian system. *Current Biology: CB*, *27*(12), 1768–1775.e3.

Wehr, T. A., Moul, D. E., Barbato, G., Giesen, H. A., Seidel, J. A., Barker, C., & Bender, C. (1993). Conservation of photoperiod-responsive mechanisms in humans. *The American Journal of Physiology, 265*(4 Pt 2), R846–R857.

Windred, D. P., Burns, A. C., Lane, J. M., Saxena, R., Rutter, M. K., Cain, S. W., & Phillips, A. J. K. (2024). Sleep regularity is a stronger predictor of mortality risk than sleep duration: A prospective cohort study. *Sleep, 47*(1). https://doi.org/10.1093/sleep/zsad253

Wong, K. Y. (2012). A retinal ganglion cell that can signal irradiance continuously for 10 hours. *The Journal of Neuroscience: The Official Journal of the Society for Neuroscience, 32*(33), 11478–11485.

Wright, Kenneth P., Jr, Andrew W. McHill, Brian R. Birks, Brandon R. Griffin, Thomas Rusterholz, and Evan D. Chinoy. 2013. Entrainment of the human circadian clock to the natural light-dark cycle. *Current Biology: CB* 23 (16): 1554.

Yetish, G., Kaplan, H., Gurven, M., Wood, B., Pontzer, H., Manger, P. R., Wilson, C., McGregor, R., & Siegel, J. M. (2015). Natural sleep and its seasonal variations in three pre-industrial societies. *Current Biology: CB, 25*(21), 2862–2868.

Youngstedt, S. D., O'Connor, P. J., Crabbe, J. B., & Dishman, R. K. (2000). The influence of acute exercise on sleep following high caffeine intake. *Physiology & Behavior, 68*(4), 563–570.

Youngstedt, Shawn D., Jeffrey A. Elliott, and Daniel F. Kripke. 2019. Human circadian phase-response curves for exercise. *The Journal of Physiology* 597 (8): 2253–68.

Zeitzer, J. M., Dijk, D. J., Kronauer, R., Brown, E., & Czeisler, C. (2000). Sensitivity of the human circadian pacemaker to nocturnal light: melatonin phase resetting and suppression. *The Journal of Physiology, 526 Pt 3*(Pt 3), 695–702.

Zhu, Gewei, Sophie Cassidy, Hugo Hiden, Simon Woodman, Michael Trenell, David A. Gunn, Michael Catt, Mark Birch-Machin, and Kirstie N. Anderson. 2021. Exploration of sleep as a specific risk factor for poor metabolic and mental health: a UK biobank study of 84,404 participants. *Nature and Science of Sleep* 13 (October):1903–12.